WHO IS THE SCIENTIST-SUBJECT?

"Esha Shah richly demonstrates how the 'ways of being and belonging' affect the 'structures' of rational cognition and how the method and philosophy of reductionism stand determined by the 'affective' scientific self. Drawing widely on the history of science, auto/biographies, psychoanalysis, and metaphysics, she shows how preference for unity and order, fascination for immortality, and love of complexity have shaped the scientific estate. The bonds of love often overcome the hegemony of gene. This is an incredible feat of feminist social epistemology and a sustained provocation to historiographical doxa."

– **Upendra Baxi**, Emeritus Professor of Law, University of Warwick, UK, and University of Delhi, India

This book explores two disparate sets of debates in the history and philosophy of the life sciences: the history of subjectivity in shaping objective science and the history of dominance of reductionism in molecular biology. It questions the dominant conception of the scientist-subject as a neo-Kantian ideal self – that is, the scientist as a unified and wilful, self-determined, self-regulated, active and autonomous, rational subject wilfully driven by social and scientific ethos – in favour of a narrative that shows how the microcosm of reductionism is sustained, adopted, questioned, or challenged in the creative struggles of the scientist-subject.

The author covers a century-long history of the concept of the gene as a series of "pioneering moments" through an engagement with life-writings of eminent scientists to show how their ways of being and belonging relate with the making of the science. The scientist-self is theorized as fundamentally a feeling, experiencing, and suffering subject split between the conscious and unconscious and constitutive of personality aspects that are emotional/psychological, "situated" (cultural and ideological), metaphysical, intersubjective, and existential at the same time.

An engaging interdisciplinary interpretation of the dominance of reductionism in genetic science, this book will be of major interest to scholars and researchers of science, history, and philosophy alike.

Esha Shah is an environmental engineer by training and a social anthropologist, historian, and philosopher of science and technology by professional choice. She is Assistant Professor with the Department of Environmental Sciences at Wageningen University, the Netherlands.

Her research interests so far have broadly concerned the history and anthropology of technology-led development in India on the divide of modernity and democracy. With this book, her research interests have expanded to philosophy of subjectivity and its relation to modes of rationality, including objectivity in science.

SCIENCE AND TECHNOLOGY STUDIES
Series Editor: Sundar Sarukkai
Manipal University

There is little doubt that science and technology are the most influential agents of global circulation of cultures. Science and technology studies (STS) is a well-established discipline that has for some time challenged simplistic understanding of science and technology by drawing on perspectives from history, philosophy, and sociology. However, an asymmetry between "Western" and "Eastern" cultures continues, not only in the production of new science and technology but also in their analysis. At the same time, these cultures, which have little contribution to the understanding of science and technology, are also becoming their dominant consumers. More importantly, science and technology are themselves being modified through interaction with the historical, cultural, and philosophical world views of the non-Western cultures, and this is creating new spaces for the interpretation and application of science and technology. This series aims to take into account these perspectives and set right this global imbalance by promoting monographs and edited volumes that analyse science and technology from multicultural and comparative perspectives.

SCIENCE AND NARRATIVES OF NATURE
East and West
Edited by Jobin M. Kanjirakkat, Gordon McOuat and Sundar Sarukkai

SCIENCE AND RELIGION
East and West
Edited by Yiftach Fehige

WHO IS THE SCIENTIST-SUBJECT?
Affective History of the Gene
Esha Shah

For more information about this series, please visit: www.routledge.com/Science-and-Technology-Studies/book-series/STS

WHO IS THE SCIENTIST-SUBJECT?

Affective History of the Gene

Esha Shah

LONDON AND NEW YORK

First published 2018 by Routledge

2 Park Square, Milton Park, Abingdon, Oxon, OX14 4RN
605 Third Avenue, New York, NY 10017

Routledge is an imprint of the Taylor & Francis Group, an informa business

First issued in paperback 2020

British Library Cataloguing-in-Publication Data
A catalogue record for this book is available from the British Library

Library of Congress Cataloging-in-Publication Data
A catalog record has been requested for this book

ISBN: 978-1-138-57033-7 (hbk)
ISBN: 978-0-367-73443-5 (pbk)

Typeset in Goudy
by Apex CoVantage, LLC

TO RAM
MY EXISTENTIAL ANCHOR

CONTENTS

PREFACE

This book is a significant departure from my earlier trajectory of work. Looking back, I cannot categorically identify a singular logic that made this shift in my research trajectory from the anthropology of development and technology (tank irrigation, GMOs, farmers' suicides, the green revolution in India) to the history and philosophy of genetic science possible. In fact, it has been a number of unrelated currents in life that culminated in the making of this book.

The most significant current was deeply personal, developing over a decade, which taught me the most important lesson so prominently discussed in this book – the irrationality, the paranoia, of knowing. During this period, I struggled with rationally knowing something so clearly and still choosing not to know because "knowing" was frightening and unbearably painful. For over almost a decade, I compulsively placed my hand in the fire, each time hoping, in vain, that I would escape the burns. And in each of those occasions a highly charged emotional situation unfolded in exactly the same way, as if the emotions also acted with the precision of Newtonian science. It took what now looks like a very long time to make the pain caused by the implied loss tolerable and thereby finally embodying the rational knowledge and learning to make different choices.

During this period "the self" became an enigma; it obstinately refused to be plastic – capable of being shaped and sculpted by the dictates of rationality and reason so fervently advocated by our modern culture. This incapacity for plasticity became yet another source of agony – why can't I see what everyone else was able to see and act accordingly? Close friends and family nearly lost confidence in my discerning abilities. And the recurring bouts of melancholia were attributed to this "irrational self" that lacked capacity for reasoned action. Paradoxically though, the fear or paranoia of knowing – the not-knowledge – led me to find refuge in studying the knowledge traditions

ix

that I would have otherwise not explored. This deep intimate struggle led me to explore psychoanalysis and existential philosophy – with an assumption that knowing how the psyche works would help relieve my suffering. It obviously didn't.

These explorations, however, became catalytic sources to the thought of this book. I originally worked with an idea of doing a biography of a woman scientist – to show whether and how knowledge stems from the deeply affective realm of the self. In my search for prototypes, I found and read Evelyn Fox Keller's iconic and inspiring biography of Barbara McClintock, which opened up an entirely new path. I initially read and critically engaged with the history of genetic science and many (auto)biographical life-writings of other molecular biologists mainly as a preparation. I chose genetic science because while working on the anthropology of GMOs in India, I had, out of curiosity, also read some aspects of the science of the gene and its modification. This random engagement with the genetic science, having started somewhere, provided an impetus to systematically study the history and philosophy of molecular biology. Bringing two parallel streams of my "undisciplined" knowledge exploration together, the chapters of this book were originally written as a series of articles critically exploring how the affective is related to the cognitive in individual lives of (genetic) scientists, while in my mind I was preparing to write a biography of a woman scientist – an idea that at the time of writing this preface I have shelved though not entirely given up. As the chapters were originally written as stand-alone articles, of which three have been published in various journals, there is some inevitable repetition of the central argument. These repetitions ironically provide the glue for the chapters on the distinct subjectivity of each scientist to be collected in the book form.

The "undisciplined" move in my research trajectory has, however, been challenging to my academic career. As in life so in career, it has not been easy to fit in. Losing my job in 2013 from the STS department at Maastricht University ironically provided the freedom to explore the history and philosophy of science. Fortunately, in November 2013 I received a two-year book-writing fellowship with the Indian Institute of Advanced Study in Shimla that has made the writing of this book possible. But, at the time of writing this preface, I have returned to my alma mater, Wageningen University in the Netherlands, working on anthropology of technology and development – far away from the history and philosophy of genetic science. Writing this book was the most stimulating, challenging, and satisfying experience, and, still, its disorderly emergence out of my career trajectory threatens it to become just an aberration. Hope not.

During this chaotic journey, I have incurred many debts. I am deeply indebted to the selection committee of the IAS Shimla for "taking the risk" and "giving a chance" to a scholar of anthropology of technology and development to work on the history and philosophy of genetic science without which this book would not have seen the light of the day. The two years I spent at the IAS Shimla were intellectually and personally the most stimulating in the past decade of my career, and for that I am grateful to the thoughtful leadership of the then director Chetan Singh and to my fellow Fellows for their enormous support and critical appreciation of the then rough and unorganized thoughts on the subject. I am especially thankful to Kaustav Chakraborty, Anindita Mukhopadhyay, Meena Alexander, Albeena Shakil, Shivani Chopra, Uma Maheshwari, Sukumar Murlidharan, Yogesh Snehi, Kundan Lal Tuteja, Anirudha Chaudhary, Udayon Misra, Rajeev Kumaramkandth, Aryak Guha, Sarvachetan Katoch, Malemngamba Meetei, Mahesh Champaklal, Sharad Deshpande, and Rajesh Joshi for providing the intellectual camaraderie and emotional friendship that made those two years the most memorable for a long time to come.

I cannot thank Gordon McOuat enough for his generous support and encouragement. When I first met him at the Manipal Centre for Philosophy and Humanities in July 2015, I was unsure about the merit of the first draft of the book just completed at that time. It was enormously encouraging that Gordon read the whole draft and provided nuanced comments that not only greatly helped improve the quality of my work but also provided the much-needed invigoration to my fragile confidence in this new venture. I also owe much intellectual debt to Upendra Baxi and Ghanshyam Shah who supported and encouraged the idea so discussed in the book from the very first moment I shared the conceptualization with them, and then on they have been untiring sources of thought-provoking inspiration. I am thankful to Andy Stirling and Wiebe Bijker for reading the first draft of the Chapter 2 that eventually got published in the journal *Minerva*. My countless thanks to Sundar Sarukkai for all the encouragement and for accepting this book for publication in the series he is editing. I consider it as my privilege. My special thanks also to the editors and publishers for granting me permission to republish three chapters (Chapter 2, a version of Chapter 5, and Chapter 6) originally published in their journals.

I want to thank my long-time close friend Himali Mehta, conversations with whom have made the journey of life an enriched experience. Thanks also to my sister Neha, brother Dhaval, and my father for always being there to steady this wobbly member of the family. Finally, it makes no sense to thank Ram – my closest friend and confidant, my critic and interlocutor, who always nurtured and mothered me, my ex-husband and now an unnameable companion traveller in the journey of life – to whom this book is dedicated. Not only that this book was conceptualized and written in constant dialogue with him but compared to what all I have learnt from him, writing this book was just a footnote.

1

INTRODUCTION

Affective history of the gene

For more than a century, scientific discourses have hailed the gene as harbouring the secret of life. *What sort of entity is the gene and how its structure and action be explained* is the question that has fundamentally driven genetic science.[1] Spanning over the first half of the twentieth century, a series of discoveries postulated that genes were located at the exact position on chromosomes in the cell nucleus constituting deoxyribonucleic acid (DNA), which was considered the carrier of biological heredity. With these discoveries, a series of highly influential claims emerged that postulated one-to-one correspondence between a gene and some developmental unit, be it a phenotypic trait, enzyme, or polypeptide. And, with the discovery of the double helix the path was paved for what the Nobel Prize–winning scientist Francis Crick called the "central dogma": *there was only a one-way flow of information from DNA to the cell.* By mid-century, the beauty and simplicity of the structure of the gene and, correspondingly, a simple resolution of the riddle of life was widely celebrated in science. By the end of the century, the reductionism of physicalism (the gene is *nothing but* a physical and chemical unit) and the central dogma (hierarchical organization of the gene to the cell) have been challenged. The term gene no longer means a fixed strand of DNA beads-on-a-chromosome-string and genes do not pass one-way information commanding the rest of the cell. Instead, it is understood that lengths of DNA distributed along chromosomes are in constant dynamic exchange with the cellular environment. A cascade of transcriptional activities involving a vast number of cellular processes and a range of regulatory, structural, splicing and editing genes, messenger and microRNA, and entities such as exons and introns are reported to be essential for the particular developmental action to take place. Some scholars call this "causal democracy" (many cellular, genetic, and epigenetic processes are causally *equally* necessary in determining the developmental outcome), others call this "many-many" problem (a vast number of genes are responsible

1

for a vast array of developmental and regulatory activities) (Griffiths 2001; Griffiths and Neumann-Held 1999). More recently, in 2012, the ENCODE (Encyclopedia of DNA Elements) project has radically redefined what is called gene.[2]

The history of the concept of the gene could also be read as the history of reductionism and challenges to reductionism (Rosenberg 2006). There are many uses of the term reductionism. Most of them ultimately refer to reductionism either as methodological practice or metaphysical or ontological view. The explanatory or methodological reductionism is often explicated in three criteria: 1) physicalism – the view that genes can be explained in terms of physics and chemistry alone; 2) hierarchical organization – genes can be broken down into smaller units and explained as a culmination of lower-level units into a higher level of hierarchy; 3) spatialization of hierarchy – hierarchy manifested in physical space and has physical sanctity.[3] Reductionism as methodological tool is considered responsible for the considerable success in molecular biology to unravel mechanisms and material constituents of biological processes. However, its conflation with the metaphysical or ontological reductionism – the view that all life processes can be causally reduced to the function of genes – has largely given rise to the problematic doctrine of genetic determinism.

"There could never be a Newton for the blade of grass," said Kant. For the anti-reductionist scientists and philosophers, Kant's dictum means that biological processes are so complex that they could never be explained in physical and chemical terms. In fact, the history of the gene century can be read as an ever-widening gap between the simplistic starting assumptions and the complexity of actual data. This sometimes raged into unresolved debates between the reductionist and anti-reductionist scientists and philosophers. The solution of such science wars is often found in the politics of representation. For instance, wondering why the nature-nurture debate (what is received on birth and what is acquired after birth) in life sciences resists resolution even after a century, Evelyn Fox Keller argues that the "unreasonable persistence of the debate is to be found in the language of particulate genes". She thinks that the language of contemporary science – which has presumably countered many dogmas of the inaugural language of the particulate genes – will "help us out of the morass" (Keller 2010, 10–13). Many scholars share Keller's views. A highly influential core of such scientific and philosophical critiques has emerged in the last quarter of the twentieth century.[4] Despite these counter-challenges, the gene side story has powerfully mutated in myriad of scientific, cultural, entrepreneurial, and popular imaginations to the extent that the gene has acquired the status of a new soul that has come to represent the essence of humanity.[5]

2

In the background of these philosophical debates on reductionism, this book asks a different question: How and why in the creative struggles of individual scientists reductionism was sustained, adopted, questioned, and challenged? The book is an attempt to reinterpret the history of reductionism in genetic science over the twentieth century as an affective/emotive history. The book discusses five pioneering moments in the history of reductionism in genetic science/molecular biology – 1) the founding of the idea of the particulate gene in the work of H. J. Muller, 2) the gene as code-script as posited in the Nobel Prize–winning physicist Erwin Schrödinger's book *What Is Life?*, 3) the discovery of the double helix in the science of Rosalind Franklin, 4) the challenge to the concept of the particulate gene in Barbara McClintock's theories on control and transposition, and 5) the hyper-reductionism of Craig Venter in sequencing and mapping the human genome. Discussing these select "contingent" moments in the history of genetic science that shaped or challenged the reductionist belief system, *I aim to posit in the book that scientists' particular ways of being and belonging pioneer the structures of rational and cognitive thought.* My claim is that intellectual paradigms are *affect worlds*, in other words, the conceptual theories are *isomorphic* with the world emotionally and existentially desired.[6] The book aims to explain the power and dominance of reductionism in science, including the challenges to this dominance, by explaining how the method and philosophy of reductionism operated in the microcosm of individual affective lives of scientists. My arguments in the book are twofold. Firstly, I argue that the scientist-self is not entirely a rational entity making deliberate theory choices based on empirical evidences. The scientist-subject is a feeling and suffering affective self, making choices in her science that means choosing her mode of existence. Secondly, I show how this affective self is profoundly constitutive of the method and philosophy of science.

The purpose of this book is therefore to philosophically engage with the question: Who is the scientist-subject? How and why scientists' particular ways of being and belonging relate to knowledge practices? Who is the scientist-self making science, who is the "actor" who is "enacting"? Chapter 2 critically reviews the work of Lorraine Daston and her co-authored work with Peter Galison on the history of scientist-subjectivity and its role in the making of objective science (Daston and Galison 2007). It also discusses Daston's later work, "On Scientific Observation", and her earlier essay, "The Moral Economy of Science" (Daston 1995, 2008). The main purpose of this chapter is to closely review these texts for their theorization of the scientist-subject either as a neo-Kantian ideal – a unified and wilful, self-determined, self-regulated, active and autonomous, rational subject wilfully driven by social and scientific ethos – or a Foucauldian construct – the subject reduced entirely to the effects of power. The main purpose of this chapter is to posit four challenges to the neo-Kantian and Foucauldian constructions of the co-implication of psychology and epistemology. Challenging Daston and Galison's argument of the co-construction of psychology/subjectivity and epistemology/objectivity, I aim to claim, following Michael Polanyi (following Sartre), that *existence precedes essence*, or, in other words, *psychology precedes epistemology*.

Developing this critique of the neo-Kantian ideal of the scientist-self, in the rest of the chapters I have explored how understanding and disclosing "entities" in scientific domain presupposes an understanding of *being* in general. Also, through the critique of the neo-Kantian scientist-self I make a case for an alternative conception of the scientist-subject and accordingly an affective and existential conception of science. My own framework of scientist-subjectivity is inspired and informed by the psychoanalytical and existential philosophies which I further historically and culturally contextualize towards *becoming* of the individual scientist-subject. Here, in this chapter, I have briefly introduced some of these ideas, which I further discuss in detail in the chapters discussing the individual affective lives of the chosen scientists. While the chapters discussing individual scientists' lives employ a variety of theoretical sources – philosophical, psychoanalytical, existential, and history of emotions and psychology of our times – in no way do I mean to suggest that the conception of scientist-subjectivity is limited to only these abstractions. While I challenge the positivist and foundationalist, especially the neo-Kantian, idea of the subjectivity, I do not boast to provide an alternative normative or regulative theory of the scientist-subjectivity. Such normative or regulative standards solidify into rigid structures while no such regulative ideal can provide an all-encompassing prescription for human condition or capacity (Flax 1990, 93). Subjectivity is conceived here as a process expressing multiple agentic qualities rather than an effect of the predetermined normativity fixed in time and space. The book tells multiple stories in a variety of styles to appreciate the complexity and multiplicity of human and hence scientific subjectivity.

The scientist-self and mode of modern science: methodology

> Selfhood is completely psychic, utterly somatic, essentially inter-corporeal and intersubjective, constantly changing . . . and funda-mentally located in space and time.
>
> (Lafrance 2007, 276)

My own project is similar to that of Lorraine Daston's and her co-author Peter Galison's work, which is closely discussed in Chapter 2, that is, to show not only that objectivity qua epistemology has history and hence it is variable but to demonstrate how it is inter-articulated with ethos and forms of subjectivity. The discussion in Chapter 2 on Daston and Galison's extensive work on the co-implication of objectivity and subjectivity raises several questions for my own idea of the scientist-subject. 1) How do psychological connect to epistemological in science? How the individual with the specific emotional and life trajectory relates to the collective. 2) What it means to *becoming* a scientist-subject from *being* and how this becoming of the scientist-subject relates to the mode of (modern) science. 3) How the making-of-science is the interplay between rationality and irrationality. 4) How objectivity in science is "super-duper" intersubjectivity (in the words of Ian Hacking quoting Kant) (Hacking 2012, 20). 5) How do the "unconscious" order or obstruct the formation of the conscious reason? In short, how does the psychic life of the scientist-subject relate to the mode of science and how do practices of science constitute subjectivity?

To ascertain adequate responses to these questions, I have adopted a range of notions such as the subject, the individual, social character, personality structure, and the self to denote the scientist-subject. These notions are not entirely interchangeable and often mean different things in different scholarly literature and still there is an underlying commonality in their use that makes it possible to discuss them in the same breath. Adopting these notions has made it possible to refer to a much larger corpus of literature, a substantial part of which has emerged from the neo-Marxist and neo-Freudian scholarly traditions in the post–World War II period; it also includes some of the recent literature on the history of emotions, individualism, narcissism, and psychology of current times. I have explained the lexicon wherever necessary, however, remaining terminologically undisciplined has proven most productive for my work.

Furthermore, I have adopted a biographical method. That is, I have read, interpreted, and related individual scientist's life-writings, biographies, and autobiographies with her life, thought, and science to frame the role of subjectivity in the making of science. The modern scientist-subject and the subjectivity of individual scientist are brought together by endorsing Thomas Mann's view that the individual consciously or unconsciously lives not just her personal life but also the life of her epoch and contemporaries.[7] The revision of the notion of the subject of science would therefore mean opening up a vista of questions that have not yet been adequately probed in the history of science. Methodologically, as already briefly discussed, I have selected five pioneering moments from the century-long history of the concept of the gene. They are 1) the founding of the concept of the particulate gene in the work of H. J. Muller, 2) the gene as code-script by

Erwin Schrödinger, 3) the gene as dynamic and movable entity as proposed by Barbara McClintock, 4) the discovery of the double helix in Rosalind Franklin's science, and 5) Craig Venter and the sequencing of the human genome. These pioneering scientists play as interlocutors for each pioneering moment. I have probed the published (auto)biographical and life-writings of these scientists in order to relate the scientists' particular ways of *being and belonging* in the world with the making of their science.

Here, a word on the methodology of reading scientific biographies as source material for history writing is pertinent. Scientific biography as a mode of doing history of science is not new. In the last decade or so a series of impressive biographies of scientists have appeared which prompts that the genre is transforming into an innovative mode of science history.[8] The history of collective thought often is nothing but a collection of a series of individual contributions – in the words of Thomas Carlyle, "history is the essence of innumerable biographies". In such histories represented purely as products of the mind, thoughts are treated as independent, self-sufficient creations, verified, falsified, or otherwise evaluated apart from the personal context out of which they arise. This way, the history of scientific thought is resolutely divorced from a scientist's life. Biographies collectively constitute serious research that give attention to the individual scientist and also link these individual lives to larger questions in the history of science. Biographies in fact shed light on the scientific collective through micro-history of the life of an individual scientist that other forms of history of science fail to accomplish.[9] However, a number of objections can be directed against biography as a primary source of history writing. Firstly, often biographies are hagiographies. It is often the case that scientific biographers, busy creating, as a historian of science describes, "scientific Cinderella" – "a hero with dazzling brilliance rising from marginality to stardom" (Abir-Am 1991, 327) – hardly pay any attention to the deeper psychological, emotional, and existential self.[10] However, this study, with the goal to cover the century-long concept of the gene and hence a number of pioneering scientists, had to depend upon the published life-writings. It is indeed a limitation of this study that it is dependent upon biographies that do not give enough attention to the intrapsychic and intersubjective life of scientists. Secondly, such objection to using biography as primary source for history writing is the attention to individual subjectivity, resulting in what Steven Shapin called "atomizing particularism" and the mind reading. Shapin suggests that the individual reflexes should be disciplined by sociologists' collectivism (Shapin 1992, 354–355). There are no easy ways to avoid these suspicions.

My defence, however, is as follows. I have derived from a number of excellent biographies and published literature on life and work of the chosen scientists. The choice of the scientist as an interlocutor for this book in fact has depended upon the availability of credible and in-depth biographical literature. Referring to multiple biographies with contrasting messages for scientists like Barbara McClintock and Rosalind Franklin has provided significant interpretative space for this book project. Also, the biographical portrait of a particular scientist discussed here – for instance, H. J. Muller, Erwin Schrödinger, and Craig Venter – is used in the background of the collective experiences in a particular historical time. The biographical literature is placed and analysed in the wider historical, institutional, political, and sociocultural contexts of the development of genetic science. To avoid mind reading, I have remained sensitive to the vocabulary the scientist has used about herself and her work; I have also given primary importance to the person who says "I" in the work, letters, drafts, sketches, personal secrets, and the metaphysical world view; I have also focused on the *enabling conditions of the self-assertions*, the search for the *emotional and metaphysical longings* of the scientist, the scientist's struggle for *continuous renewal of oneself*, and his or her quest for *personal authenticity*; I have also focused on the sources of suffering, fears, anxiety, and phobias whenever such details are available; and lastly, I have asked and prioritized questions such as: *What was the kind of world that the scientist sought to realize? What was the scientist's emotional or metaphysical a priori?*[11]

In positing the scientist-self as a fundamentally feeling, suffering, and experiencing subject, I have theorized the unfolding of the scientist-self as constitutive of the following aspects – 1) emotional/psychological/affective as "situated" (cultural and ideological), 2) metaphysical, 3) intersubjective, and 4) existential. These aspects are not mutually exclusive or even easily separable. The following list is discussed only for the analytical convenience. I must also emphasize that not all these theoretical sources of the self are used for each and every chapter. They are selectively used for the specific arguments in each chapter towards painting a truly multiple subjectivity of the scientist-self.

Emotional/psychological/affective as situated: I have located the individual scientist-self in a particular historical, cultural, and ideological context and have ascertained what Fernand Braudel called *mentalités* of the generation. I have also selectively derived from the growing literature on history of emotions, sources of self, and the theories on socio-cultural-psychology of those times. Affects and emotions here are interchangeably used and meant to provide supra-individual and intersubjective meaning to the historical present and life experiences. The emphasis is on understanding how the emotional investments circulate through the cultural landscape and how they lead the way for belonging and shape the rational thought and normative workings of scientific institutions.[12]

Metaphysical: Science cannot progress without the powerful assumptions about the world it is trying to investigate. What metaphysics does the scientist implicitly or explicitly embrace? I have particularly focused on the metaphysical search for order, unity (Chapter 4 on Erwin Schrödinger), immortality (Chapter 3 on H. J. Muller), and complexity (Chapter 5 on Barbara McClintock) and the way in which these quests have shaped the individual scientific personalities and culminated into the belief systems of science.

Objectivity as intersubjectivity: Ian Hacking explains how Kant described objective and subjective, "practical principles . . . are subjective . . . when the condition is regarded by the subject as valid only for his own will. They are objective . . . when they are recognized as . . . valid for the will of every rational being". Hacking thus defines "objective as inter-subjective", or rather, "super-duper obligatory inter-subjectivity" (Hacking 2012, 20). The process of crystallizing objectivity is thus intensely intersubjective. These intersubjective dramas involve the "collectivism of affect-saturated values" as Lorraine Daston argues (Daston 1995, 4); they also consist of intense entanglement with other subjects and with objects of science, including power struggles. Especially in the chapters on Barbara McClintock, Craig Venter, and Rosalind Franklin, I have followed the agency of the scientist as it emerges or unfolds during the process of *intra-action*[13] – entanglement with other epistemic agents, apparatuses, epistemic concepts, and the object of knowledge – in the process of the making of objective science.

Existential: I have hermeneutically interpreted existential drives of the scientists which cannot be reduced to his or her cultural or historical situatedness. As Polanyi (quoting Sartre) posited *existence precedes essence*, science involves a series of existential choices. Several philosophies on existentialism since Kierkegaard have pointed out that the individual not only knows oneself but actually "chooses oneself" in a sense that the individual is determined by his or her own necessity and authenticity (Söderqvist 1996, 72–74). The question of choice and freedom are crucial here. In terms of Lacan interpreted by Žižek, the subject only retroactively posits the causes of its own existence, the freedom is also posited only retroactively, in the sense that the subject does not select from an infinite list of possibilities but chooses the "necessities that will determine himself".[14] For Žižek this is the Lacanian paradox of the subject, a parallax moment, the choice that is no choice, or the choice that is nothing but choosing the necessity of one's own existence and hence no choice (Žižek 2006). How is science an expression of the deeper existential choices for the scientist-subject? This question is foregrounded in all chapters.

Psychoanalytical and *becoming* from *being*: Several chapters have adopted a psychoanalytical approach. Following feminist thinkers on psychoanalysis, the chapter on Barbara McClintock (Chapter 5) relies especially on the branch of psychoanalytical theory concerned with the development of self in relation to others, called object relations theory as in the work of Donald Winnicott (1958, 1965) also discussed by the feminist scholars (Benjamin 1988, Flax 1990). Object relations theory permits to understand the ways in which the earliest experiences structured by socially determined family relationships shape the subject's conception of the world and relation to other objects and subjects. In building my theory of objectivity-subjectivity I have kept in centre the most important aspect of humanity's relationship with the modern science, that is, the desire for mastery and control of nature. There are two philosophical traditions that engage with this aspect – one relates to the birth and another with the death. Since the pioneering work of scholars of the Frankfurt School there have been several attempts to explain the psychic roots of mastery and control of nature in the psychodynamics of narcissism. Another such scholarly tradition emerges from the existential philosophies on human finiteness. Both these philosophical traditions posit the nature of *being*. In Chapter 5, the chapter on Barbara McClintock, I have further discussed the neo-Freudian thinking on narcissism and intersubjectivity, and in the chapters on H. J. Muller I have discussed the existential philosophies on human immortality. I aim to show how these states of *being* are sources of mastery and control of nature and hence the existential source of the advance of science. Further, my own framework of the scientist-self emerges from relating these theories on the idea of *being* with the processes of *becoming* – the changing personality, character, or social history of the individual over the twentieth century. This way I aim to combine the philosophical traditions on *being* with the sources of historical literature on "individuality". Chapter 7 on Craig Venter discusses the formation of contemporary personality in the context of shifting "psychology of our times". In this chapter I have sketched a framework for the historically changing character of the scientist-self that includes the existential debates.

Notes

1 I have used the term genetic science to combine both classical genetics – that emerged since the rediscovery of the Mendelian theory of heredity at the beginning of the twentieth century – and molecular genetics (biology) – that was organized since the discovery of the double helix in the mid-twentieth century. For further discussion on the disciplinary distinctions in life sciences and genetics, see (Kitcher 1984; Powell et al. Forthcoming).

2 This is only a brief description of a long history of the concept of the gene. This will be further discussed in the individual chapters. Here, I have referred to a number of excellent histories of molecular biology: Olby (1994), Kay (2000), Portugal and Cohen (1977), Kay (1993), Schwartz (2008), Judson (1979). For a historical analysis of reductionism and biological determinism in life sciences, see Rose (1997), Lewontin (1995), and Fuerst (1982); for the passionate defence of reductionism in molecular biology, see Rosenberg (2006); for the history of rhetorical transformation in molecular biology, see Doyle (1997); for the history of the century-long concept of the gene, see Keller (2002, 1991, 1990); for the history of the gene as a boundary object for the molecular biology, see Rheinberger (2000); for the history and philosophy of genomics, see Zwart (2009); for the challenge to the gene as a reductionist explanatory framework, see Griffiths and Neumann-Held (1999), Keller (2000), and Griffiths (2001); and for the concept of "causal democracy" and the further challenge to the idea of the gene, see Griffiths and Gray (1994).

3 For further discussion, see Sarkar (1992, 1998, 1996) and Beckwith (1996).

4 In addition to the references cited in footnote 5, for the philosophical critique of molecular biology, see Dupre (1993), Griffiths (1992), and Rheinberger and Gaudilliere (2004).

5 For instance, Ian Hacking shows how the "genetic imperative" – the drive to find genetic underpinnings to all things human – is creating new forms of biosocial identities (Hacking 2006). For the discussion on powers of DNA in popular culture, see Nelkin and Lindee (2004). For the gene as the essence of humanity, see Ridley (1999), Mauron (2001), and Dawkins (2006), and for how the gene is represented in popular culture like cinema, see Stacey (2012).

6 The concept of isomorphism between the affective and intellectual paradigm is referred to here from the work of Thomas Söderqvist's biography of the Nobel Prize–winning scientist Niels Jerne. Söderqvist further refers to the earlier scholarly work of Lewis Feuer and asks what is the *emotional a priori* of the scientist, what kind of world that the scientist on emotional grounds sought to realize in the scientific theorizing (Söderqvist 1996, 69).

7 As quoted in Söderqvist (1996, 50).

8 See, for the defence of biographies as a mode of scientific history, Nye (2006) and Söderqvist (1996, 69, 2011).

9 For example, in the fascinating biography of Darwin, the biographers show how the hatred of slavery shaped Darwin's views on human evolution (Desmond and Moore 1991). Yet another example is the way Katherine Hayles uses Norbert Wiener's autobiography and argues that the scientist's tormented hallucinations were integral to his science of cybernetics. She thereby challenges the idea of the scientist-self as primarily a rational and cognitive agent (Hayles 1999, 84–112). Further, in one of the rare studies that focuses on sexuality and science, Elizabeth Wilson uses biographical literature to show the way affect, sexuality, and brilliant scientific work overlapped in the early history of Artificial Intelligence (Wilson 2009).

10 For a detailed review of several scientific biographies of molecular biologists, see Abir-Am (1991).

11 For the further discussion on the existential conception of scientific biography, see Söderqvist (1996).

12 The literature on the modern age of personality, individualization, possessive individualism, and psychology of our times is generally referred. For example, see, Gauchet (2000). Other theories of the self, such as the seminal work of Charles Taylor on historical sources of what it means to be a human agent in the modern world and Christopher Lasch's use of the psychoanalytic theories to understand the emergence of what he calls the culture of narcissism in the post–World War II period is referred (Taylor 1989; Lasch 1979). The burgeoning literature on the "affective turn" in social sciences is carefully processed and consulted; however, I do not make a distinction between emotions, feelings, and affects, as much of this literature does. See, for example, Massumi (2002), Blackman and Venn (2010), Clough (2010), Clough and Halley (2007), and Pile (2010). For the critique of the affective turn, see Hemmings (2005) and Tyler (2008).

13 Karen Barad defines intra-action as mutual constitution of entangled agencies. This is in contrast to the usual "interaction", which assumes the existence of separate individual agencies that precede their interaction. The notion of intra-action means that distinct entities do not precede but emerge through the intra-action (Barad 2006, 33).

14 As discussed in Söderqvist (1996, 72–74).

References

Abir-Am, Pnina. 1991. "Nobelesse Oblige: Lives of Molecular Biologists." *Isis* 82 (2):326–343.

Barad, Karen. 2006. *Meeting the Universe Halfway: Quantum Physics and the Entanglement of Matter and Meaning*. Durham: Duke University Press.

Beckwith, Jon. 1996. "The Hegemony of the Gene: Reductionism in Molecular Biology." In *The Philosophy and History of Molecular Biology: New Perspectives*, edited by Sahotra Sarkar, 171–184. Dordrecht: Kluwer Academic Publishers.

Benjamin, Jessica. 1988. *The Bonds of Love: Psychoanalysis, Feminism, and the Problem of Domination*. New York: Pantheon Books.

Blackman, Lisa, and Couze Venn. 2010. "Affect." *Body and Society* 16 (1):7–28.

Clough, Patricia. 2010. "Afterword: The Future of Affect Studies." *Body and Society* 16 (1):222–230.

Clough, Patricia, and Jean Halley, eds. 2007. *The Affective Turn: Theorizing the Social*. Durham and London: Duke University Press.

Daston, Lorraine. 1995. "The Moral Economy of Science." *Osiris* 10 (2nd Series):2–24.

Daston, Lorraine. 2008. "On Scientific Observation." *Isis* 99 (1):97–110.

Daston, Lorraine, and Peter Galison. 2007. *Objectivity*. Brooklyn: Zone Books.

Dawkins, Richard. 2006. *The Selfish Gene*. New York: Oxford University Press.

Desmond, Adrian, and James Moore. 1991. *The Life of a Tormented Evolutionist: Darwin*. New York: A Time Warner Company.

Doyle, Richard. 1997. *On Beyond Living: Rhetorical Transformations of the Life Sciences*. Stanford: Stanford University Press.

Dupre, John. 1993. *The Disorder of Things: Metaphysical Foundations of the Disunity of Science*. Cambridge, MA: Harvard University Press.

Flax, Jane. 1990. *Thinking Fragments: Psychoanalysis, Feminism, and Postmodernism in the Contemporary West*. Berkeley: University of California Press.

Fuerst, John. 1982. "The Role of Reductionism in the Development of Molecular Biology: Peripheral or Central?" *Social Studies of Science* 12 (2):241–278.

Gauchet, Marcel. 2000. "A New Age of Personality: An Essay on the Psychology of Our Times." *Thesis Eleven* 60:23–41.

Griffiths, Paul, ed. 1992. *Trees of Life: Essays in Philosophy of Biology*. Dordrecht: Kluwer Academic Publishers.

Griffiths, Paul. 2001. "Genetic Information: A Metaphor in Search of a Theory." *Philosophy of Science* 67:26–44.

Griffiths, Paul, and R. D. Gray. 1994. "Developmental Systems and Evolutionary Explanation." *The Journal of Philosophy* 91 (6):277–304.

Griffiths, Paul, and Eva Neumann-Held. 1999. "The Many Faces of the Gene." *BioScience* 49 (8):656–662.

Hacking, Ian. 2006. "Genetics, Biosocial Groups and the Future of Identity." *Daedalus* (Fall):81–95.

Hacking, Ian. 2012. "Objectivity in Historical Perspective." *Metascience* 11:11–39.

Hayles, Katherine. 1999. *How We Became Posthuman*. Chicago: University of Chicago Press.

Hemmings, Clare. 2005. "Invoking Affect: Cultural Theory and the Ontological Turn." *Cultural Studies* 19 (5):548–567.

Judson, Horace. 1979. *The Eighth Day of Creation: Makers of the Revolution in Biology*. New York: Simon & Schuster.

Kant, Immanuel. 1987. *Critique of Judgment*. Translated with an Introduction by Werners Pluhar. Hackett Publishing Company, Indianapolis/Cambridge, p. 311.

Kay, Lily. 1993. *The Molecular Vision of Life: Caltech, the Rockefeller Foundation, and the Rise of the New Biology*. Oxford: Oxford University Press.

Kay, Lily. 2000. *Who Wrote the Book of Life? A History of the Genetic Code*. Stanford: Stanford University Press.

Keller, Evelyn Fox. 1985. *Reflections on Gender and Science*. New Haven: Yale University Press.

Keller, Evelyn Fox. 1990. "Physics and Emergence of Molecular Biology: A History of Cognitive and Political Synergy." *Journal of the History of Biology* 23 (3):389–409.

Keller, Evelyn Fox. 1991. "Fractured Images of Science, Language, and Power: A Postmodern Optic, or Just Bad Eyesight?" *Poetics Today* 12 (2):227–243.

Keller, Evelyn Fox. 2000. *The Century of the Gene*. Cambridge: Harvard University Press.

Keller, Evelyn Fox. 2002. *Making Sense of Life: Explaining Biological Development with Models, Metaphors, and Machines*. Cambridge, MA: Harvard University Press.

Keller, Evelyn Fox. 2010. *The Mirage of a Space between Nature and Nurture*. Durham: Duke University Press.

Kitcher, Philip. 1984. "1954 and All That: A Tale of Two Sciences." *The Philosophical Review* 93 (3):335–373.

Lafrance, Marc. 2007. "Embodying the Subject: Feminist Theory Meets Contemporary Clinical Psychoanalysis." *Feminist Theory* 8 (3):263–278.

Lasch, Christopher. 1979. *The Culture of Narcissism*. New York: Norton.

Lewontin, R. C. 1995. *Biology as Ideology: The Doctrine of DNA*. New York: HarperCollins Publishers.

Massumi, Brian. 2002. *Parables of the Virtual: Movement, Affect, Sensation*. London: Duke University Press.

Mauron, Alex. 2001. "Is the Genome the Secular Equivalent of the Soul?" *Science* 291 (5505):831–832.

Nelkin, Dorothy, and M. Susan Lindee. 2004. *The DNA Mystique: The Gene as a Cultural Icon*. Ann Arbor: University of Michigan Press.

Nye, Mary Jo. 2006. "Scientific Biography: History of Science by Another Means?" *Isis* 97 (2):322–329.

Olby, Robert. 1994. *The Path to the Double Helix: The Discovery of DNA*. New York: Dover Publications, Inc.

Pile, Steve. 2010. "Emotions and Affect in Recent Human Geography." *Transactions of the Institute of British Geographers* 35 (1):5–20.

Portugal, F. H., and J. S. Cohen. 1977. *A Century of DNA: A History of the Discovery of the Structure and Function of the Genetic Substance*. Cambridge, MA: The MIT Press.

Powell, Alexander, Maureen O'Malley, Staffan Müller-Wille, Jane Calvert, and John Dupre. Forthcoming. "Disciplinary Baptisms: A Comparison of the Naming Stories of Genetics, Molecular Biology, Genomics and Systems Biology." *History and Philosophy of the Life Sciences* 29 (1):1–23.

Rheinberger, Hans-Jorg. 2000. "Gene Concepts: Fragments from the Perspectives of Molecular Biology." In *The Concept of the Gene in Development and Evolution: Historical and Epistemological Perspectives*, edited by Peter Beutron, Raphael Falk and Hans-Jorg Rheinberger, 219–239. Cambridge: Cambridge University Press.

Rheinberger, Hans-Jorg, and Jean-Paul Gaudilliere, eds. 2004. *Classical Genetic Research and its Legacy: The Mapping Culture of Twentieth-Century Genetics*. London: Routledge.

Ridley, Matt. 1999. *Genome: The Autobiography of Species in 23 Chapters*. New York: Harper Collins.

Rose, Steven. 1997. *Lifelines: Biology, Freedom, Determinism*. London: Penguin Books.

Rosenberg, Alex. 2006. *Darwinian Reductionism: How to Stop Worrying and Love Molecular Biology*. Chicago: University of Chicago Press.

Sarkar, Sahotra. 1992. "Models of Reduction and Categories of Reductionism." *Synthese* 91 (3):167–194.

Sarkar, Sahotra. 1996. "Biological Information: A Skeptical Look at Some Central Dogmas of Molecular Biology." In *The Philosophy and History of Molecular Biology*, edited by Sahotra Sarkar, 187–232. Dordrecht: Kluwer Academic Publisher.

Sarkar, Sahotra. 1998. *Genetics and Reductionism*. Cambridge: Cambridge University Press.

Schwartz, James. 2008. *In Pursuit of the Gene: From Darwin to DNA*. Cambridge, MA: Harvard University Press.

Shapin, Steven. 1992. "Discipline and Bounding: The History and Sociology of Science as Seen through the Externalism-Internalism Debate." *History of Science* 30:333–369.

Söderqvist, Thomas. 1996. "Existential Projects and Existential Choice in Science: Science Biography as an Edifying Genre." In *Telling Lives in Science: Essays in Scientific Biography*, edited by Michael Shortland and Richard Yeo, 45–84. New York: Cambridge University Press, 54.

Söderqvist, Thomas. 2011. "The Seven Sisters: Subgenres of *Bioi* of Contemporary Life Scientists." *Journal of the History of Biology* 44:633–650.

Stacey, Jackie. 2012. *The Cinematic Life of the Gene*. Durham, NC, and London: Duke University Press.

Taylor, Charles. 1989. *Sources of Self: The Making of the Modern Identity*. Cambridge: Harvard University Press.

Tyler, Imogen. 2008. "Methodological Fatigue and the Politics of the Affective Turn." *Feminist Media Studies* 8 (1):85–90.

Wilson, Elizabeth. 2009. "'Would I Had Him with Me Always': Affects of Longing in Early Artificial Intelligence." *Isis* 100 (4):839–847.

Winnicott, Donald. 1958. "The Capacity to be Alone." *The International Journal of Psychoanalysis* 39:416–420.

Winnicott, Donald. 1965. "New Light on Children's Thinking." In *Psychoanalytic Explorations*, edited by Donald Winnicott, 152–157. Cambridge: Harvard University Press.

Žižek, Slavoj. 2006. *The Parallax View*. Cambridge, MA: The MIT Press.

Zwart, Hub. 2009. "The Adoption of a Map: Reflections on a Genome Metaphor." *Genomics, Society, and Policy* 5 (3):29–43.

2

WHO IS THE SCIENTIST-SUBJECT?

A critique of the neo-Kantian scientist-subject in Lorraine Daston and Peter Galison's *Objectivity*[1]

> My parents, grandparents, great-grandparents, my brothers and sisters, and all my family, and those close to me, people who are my friends, those intimately near to me, as well as my comrades and my acquaintances, together with all those more distant people who, during these last years, have been around me, that is to say army people, employees, my financial and worldly relations, have most often deceived, insulted, scorned, railed, mocked, mumbled, dishonoured, brutalised, thrashed me.[2]

These were the words of the well-known mathematician Georg Cantor who often suffered psychotic episodes and was hospitalized many times during the development of his revolutionary set theories – now known as Cantorian set theories foundationally important in mathematics and science (Burgoyne 2002, 237). Psychoanalyst Imre Hermann, who studied the creative work of Cantor, argued that this was actually an example of parallelism between the psychic structure of Cantor's manic-depressive episodes and the structure of his set theory.[3] If one closely looks at the structure of the statement, there are two types of entities here – the people and Cantor's perception of what these people are doing to him. Cantor's relationship with the first entity becomes distant as the list progresses – begins with parents and ends with the general worldly relations. But in the list of the second entity, Cantor's perception of these people's actions becomes more and more physically intense; beginning with being deceived, Cantor feels brutalized and thrashed as the list progresses – the stranger the person the more brutalizing the impact. This is the core of the set theory in mathematics: how the entities in one set relate to the entities in the other, especially when both sets encompass the infinity of members.

"What are the psychic pre-requisites for the advance of science? How the analysis of the soul is structured vis-á-vis the analysis of science?" These were the questions Freud was grappling with in his last work *Moses and Monotheism* published in 1938. The same question is asked differently in recent times. "Is Science emotional?" asks White (2009). Such a question about science might even mean a heresy. The view that science, supposed to be objective, depends in essential ways upon highly specific constellations of emotional and existential – subjective – experiences is fundamentally paradoxical. In fact, subjectivity and objectivity are commonly arrayed in opposition. More than often, objectivity is a denial, wilful control, or erasure of subjectivity. Steven Shapin calls it a dustbin conception of subjectivity – the bin collects those stories that deflate, disrupt, or disorder objectivity (Shapin 2012, 171–172). As Keller argues, the dream of completely objective science contains images of self as autonomous and objectified, severed not only from the outside world of other objects and subjects but also from their own subjectivity (Keller 1985, 150). Not only that subjectivity disrupts objectivity, but the role of individual scientists' subjectivity in the making of science is often obliterated, treated as merely incidental, designated as mind reading, or relegated to biographies. Even when subjectivity is accounted as co-implicated with objectivity, it is treated only as mental states of collectives. Lorraine Daston discusses the way certain methodological paradigms such as empiricism and objectivity require *moral economies* of *affect-saturated values* that define mental states of collectives (Daston 1995, 2008). In their historical analysis of the co-production of subjectivity-objectivity, Daston and Galison insist that they were concerned with the collectives and not with the individual psychology (Daston and Galison 2007). Steven Shapin also warns against the risk of "atomizing particularism" and suggests that the individual reflexes should be disciplined by sociologists' collectivism (Shapin 1992, 354–355). And when the scientist's subjectivity is indeed treated in individual capacity, it is predominantly constructed as the neo-Kantian ideal – a unified and wilful, self-determined, self-regulated, active and autonomous, rational subject wilfully driven by social and scientific ethos. Another approach popular among historians is the scientist-subject as a Foucauldian construct who is reduced to the effects of power. Cantor is the kind of example that historians of science are more likely to treat as a genius scientist gone mad with whom they would be wary of engaging outside the biographical pages. Taking the example of Cantor, I wish to modify White's question "Is science emotional?" into "Who is the scientist-subject doing science?" In fact, this question is being increasingly engaged with in the recent works in history of science.[4] And still, it would be hard to find the "mad excess" of Cantor, especially his delirium and his suffering, treated as not just integral but as an indispensable part of his science.

16

The purpose of this chapter is to closely engage with the work of Lorraine Daston and her co-authored work with Peter Galison on the history of scientist-subjectivity in the making of objective science. The chapter mainly engages with Daston and Galison's book *Objectivity* (Daston and Galison 2007) while it also discusses Daston's later work "On Scientific Observation", and her earlier essay "The Moral Economy of Science" (Daston 1995, 2008). The main purpose of this chapter is to posit four challenges to the neo-Kantian and Foucauldian constructions of the co-implication of psychology and epistemology presented in these texts. Challenging Daston and Galison's argument of the co-construction of psychology/subjectivity and epistemology/objectivity, I want to propose that understanding and disclosing "entities" in scientific domain presupposes an understanding of "being" in general. By critiquing the neo-Kantian scientist-self I wish to make a case for an alternative conception of the scientist-subject and accordingly an affective and existential conception of science. I wish to claim, following Michael Polanyi (following Sartre), that *existence precedes essence*, or in other words, psychology precedes epistemology.

The scientist-self in *Objectivity*

In their book *Objectivity*, Lorraine Daston and Peter Galison discuss a rich array of scientific atlases as records of the co-construction of objectivity-subjectivity. A scientific atlas is a compendium or a catalogue of visual and other images. The authors show how the atlas standardized and recorded ideal practices of both selfhood and objective science. They discuss three different forms of epistemological practices – truth-to-nature, mechanical objectivity, and trained judgment in the making of science represented by these atlases. These epistemological virtues and the corresponding notions of subjectivity, although chronologically unfolded in succession from the eighteenth to the twentieth centuries, were not mutually exclusive. Each successive stage built upon as well as reacted to the earlier ones. The main argument of the book is anchored on the mechanical objectivity of the mid-nineteenth to early twentieth centuries. Scientists in these atlases are portrayed as self-disciplined, steel-willed, even self-denying individuals who are motivated by a strong moral drive. The core task of the scientist-self in the second half of the nineteenth century was to diminish by strength of will the subjective in the making of the objective. This would mean expelling all forms of prejudice, even skill, fantasy, emotional attachment, and judgment, in order to make the cognitive and the perceptual not only "verifiable, epistemologically warranted and communicable" but also the "exact" and hence the "objective" replica of nature. The scientist-self corresponding to truth-to-nature objectivity was a sage, whereas the one who responded to mechanical objectivity was the indefatigable worker, and the one who practiced trained judgment as objectivity was the intuitive expert (Daston and Galison 2007, 3–7).

The central claim of Daston and Galison's *Objectivity* therefore is not only to show that objectivity has history and hence it is variable, but more importantly to demonstrate how objectivity – or, in other words, epistemology – is *always* inter-articulated with ethos and forms of subjectivity. What is important to note here is that the virtue of mechanical objectivity is asserted only by the suppression or expulsion of subjective forces, which is achieved by exercising extreme efforts of will – this wilfulness forms the core of the objective scientist-self. In fact, Daston and Galison compare the wilfulness with asceticism. They claim that "objectivity is to epistemology what extreme asceticism is to morality" (Daston and Galison 2007, 374). They further clarify that the demands of objectivity compel the knower to develop the most strenuous forms of self-cultivation, which even borders on the brink of self-denial and self-destruction (Daston and Galison 2007, 40). Daston and Galison thus insist on a strong link between epistemology and ethics; they argue that in pursuit of objectivity scientists convert themselves to another style of life like ascetics or religious philosophers of the antiquity. In short, in making a case that the changes in subjectivity are co-implicated with the changes in objectivity, Daston and Galison construct the scientific self as a neo-Kantian ideal – a determinate, regulated, active, and autonomous subject driven by wilful ascetic ethos.

This wilful self, however, is acknowledged to constitute fear. In her article "On Scientific Observation", which followed the publication of *Objectivity*, Daston argues that the fear that the subjective lenses might filter or distort objective empirical results has always remained in the background (Daston 2008, 97–98). Fear as constitutive of the scientist-self has a much stronger place in the book *Objectivity*. "In all cases, it is fear that drives epistemology", write Daston and Galison (2007, 49). All epistemology begins in fear – the fear that "the world is too complex or the human intellect is too weak to grasp this complexity" (Daston and Galison 2007, 49). Objectivity is one such chapter in this history of intellectual fear. Daston and Galison, however, contend that the fear that objectivity addresses is different from and deeper than the others. The threat to objectivity is not external, it is rather internal – "*Objectivity fears subjectivity, the core self*" (Daston and Galison 2007, 372–374).

As one reviewer argues, if the fear, not ethos but pathos drive epistemology, then causality is introduced between psychology and epistemology – in the sense that psychology/subjectivity and epistemology/objectivity are no longer co-constructed but psychology precedes by driving epistemology (Anderson 2008, 660–661). A number of reviewers raised similar issues with the book *Objectivity*. For example, one reviewer asked: "Is Objectivity an emanation of more fundamental subjectivity, a subjectivity that is afraid of error, of the inaccuracies that ensue from any act of willfulness, afraid ultimately of itself?" (Switzer 2009, 100–101). Paul White called it an out-break of "epistemic fear" on the pages of *Victorian Studies* (the journal in which this debate on the book took place) (White 2009, 796).

In their response to the reviewers, Daston and Galison make a case that their conception of the way the knower relates to the knowledge is *orthogonal* (italics mine) to the familiar psychological and sociological ways to understand this. They interpret psychological as pertaining to the knower as an individual with a specific emotional and intellectual life trajectory, whereas sociological as relating to the knower as a member of a collective. They theorize their own conception of the fear as a kind of "hybrid monster" that straddles the boundaries between psychological and socio-logical, and psychological and epistemological. They insist that the fear is genuinely epistemic. It is a "response to genuine and multiple obstacles to the acquisition of knowledge" (Daston and Galison 2008, 671). The fear is an "affective state" not incompatible with epistemology. They say, "episte-mology is rooted in an ethos which is at once normative and affective – or *affective because normative*" (Daston and Galison 2008, 671, italics mine). In short, for Daston and Galison, the epistemic fear is psychological, collec-tive, and epistemological at the same time, and hence it is neither patho-logical nor does its occurrence foreground psychology prior to epistemology as one of the reviewers contends. The key aspect of Daston and Galison's insistence is that the fear is "not irrational", in fact, it is "reasonable", they argue. This response would indeed keep intact the neo-Kantian sense of the self. The self that is afraid of itself is so only in the noble pursuit of objec-tivity. This self, first and foremost, is normative; it is determined to do the right thing; it is affective (fearful) only because it is normative. Active self-consciousness is thus the ground upon which both the ethics of knowledge and knowledge are built.

Later, I aim to posit four challenges to Daston and Galison's cen-tral argument on how epistemology and ethos fuse. I wish to argue that the (epistemic) fear is not rational but "irrational"; it is not ethical but "pathological". I must point out that both these terms – irrational and pathological – have specific meanings that I will clarify and discuss later in the chapter. The first challenge is derived from Jacques Lacan's argument that the subject of science constituted by the mode of modern science suf-fers from "paranoia". The second challenge builds upon Kant's own denial that the perfect correspondence between the moral law and the human will is possible. Kant in fact thought that an ethical human act is impossible without a "pathological" component. The third challenge questions the way Daston and Galison have taken appearance for being in their application of the Foucauldian concept of *technologies of the self* to constructing the scientist-subject of mechanical objectivity. And the fourth challenge contests the notion of the unconscious, especially in Daston's earlier and later work.

Impossibility of knowing the real causes paranoia

For the first challenge, I want to discuss the example that Daston and Galison open their book with. The British physicist Arthur Worthington is working on the photographic images taken a few thousandths of a second apart of impact caused by a liquid drop. He is working on a compendium of hand-drawn sketches of droplet images since 1875 that eventually contributed to the branch of fluid mechanics. For Worthington, the splash of milk droplet hitting a hard surface or dropping into a liquid break into a "perfect symmetry" (Daston and Galison 2007, 11). He ignores the accidental specificity and peculiarity of what he thinks is a defective splash and rather sets out to capture the world in its "types and regularities". The retina impressions of perfection and symmetry, however, were shattered once he captured these images for the first time in photographs in 1884. The photographs showed irregularities instead of symmetries. Not only that the imperfection of nature was shocking but what Worthington found disturbing was the realization that he was deceived of idealized, symmetrical mirages for 21 years. He was concerned that even when he did encounter many irregular and unsymmetrical images, in compiling these images for the compendium, his mind chose only the "ideal" and symmetrical splashes and rejected irregularities. Worthington concluded that the selection of idealized images was in fact the error of judgment of a fallible human. Daston and Galison argue that Worthington was not alone in rejecting the irregular. The choice of perfect over imperfect was profoundly entrenched in the scientific practices at that time. With the mechanical camera replacing the mind's eye in 1895, the earlier epistemological virtue of documenting perfect and regular images of nature was reduced to a psychological fault, a defect in perception (Daston and Galison 2007, 13–16).

Worthington's realization has an important character. The transition from the subjective and hence fallible human judgment to the mechanically produced, infallible perception meant a major shift in the way nature was visualized. From being symmetrical regularity, the physical world now acquired asymmetrical individuality and as a result became profoundly complex. In Worthington's terminology, the objective view – the new epistemic virtue – was required to capture this "real" as opposed to the subjectively produced "imaginary" nature. Daston and Galison argue that mechanical objectivity inaugurated new subjectivity that was required to wilfully reject all human intervention, all subjective elements, in the representation of nature. In projecting this new objectivity-subjectivity, however, Daston and Galison underplay the shift in the scientist's attitude towards nature. Objectivity now meant a "faithful", "real", "exact", "as it was" representation of nature. The purpose of objectivity, including the mechanical devices like microscopes and telescopes through which it was ascertained, was to bring the "remote" and "inaccessible" nature closer. The image documented by the previous, truth-to-nature epistemic objectivity, was idealized, perfected, and it was a characteristic specimen but not the exact nature. Mechanical objectivity thus made the perfected nature imperfect and instead aspired to present the object of nature "just as it was" (Daston and Galison 2007, 36, 44–45). In transiting from truth-to-nature to mechanical objectivity, nature became variable, accidental, impure, and unknown – but also real and inaccessible (Daston and Galison 2007, 51).

However, expelling everything that was subjective, and capturing the real and exact nature created anxiety among scientists. It was often expressed that representing the "real" nature in all its complexity was impossible. For example, an anatomist, Albrecht van Haller, complained that "the infinite labor was required to trace the labyrinthine variety of the arteries, which even numerous dissections had failed to coalesce into a clear pattern" (Daston and Galison 2007, 81). How to ensure what was represented was indeed a real representation of nature? Would expelling all subjectivity be enough to ensure it?

What is important to note here is that in selecting the typical and ideal images while pursuing the epistemic objectivity of truth-to-nature, scientists searched for the regularity of "God's law", the perfection of which can then be admired. Mechanical objectivity did not provide any such reference point. In fact, in mechanical objectivity, objectivity and its pursuit remained "an elusive goal, a destination always just past the horizon" (Daston and Galison 2007, 189). The impossibility of objectivity was also reflected in the impossibility of expelling all subjectivity. A naturalist, Rudolf Virchow, declared that even after trying for 30 years he was not able to de-subjectivize him entirely (Daston and Galison 2007, 189). These scientists struggled with "incoherent scientific objects" (Daston and Galison 2007, 236) and the frustration of falling short of the highest epistemic idea of objectivity was widespread (Daston and Galison 2007, 233). As a result, both objectivity and the scientific self that practiced it were declared as "intrinsically unstable" (Daston and Galison 2007, 250). On the account of this impossibility of ensuring correspondence of their representations to nature as real, "seeing nature as it is", scientists became, in Daston and Galison's own words, "well-nigh maniacal" (Daston and Galison 2007, 66). For the truth-to-nature epistemic virtue, the reference point of the perfection of God's law granted the truth of knowledge. But there is no such reference point available in mechanical objectivity to grant such certainty of knowledge. How did this shift in the attitude towards nature – the mode of modern science – constitute the subject?

It was Alexandre Koyre who argued that the birth of modern science is concomitant with the transformation of philosophical attitude; it marked the reversal of the value attached to intellectual knowledge in comparison to sensible experience. Lacan closely followed Koyre's conviction that it was Descartes who formulated the principles of new science marking the transformation in philosophical attitude, which was attributed to Cartesian *Cogito* in the form of the belief in human rationality. Following Koyre, Lacan said, "For science, the *Cogito* marks the break with every assurance conditioned by intuition" (Lacan 1995 [1964], 261). Lacan thus dubbed the Cogito endowed with capacity for reasoning as "subject of science".[5] This, Lacan describes as "a certain moment of the subject that I consider to be an essential correlate of science . . . the moment Descartes inaugurates that goes by the name *Cogito*" (Lacan 1965, 4). Lacan further states, "the modification of our subject position . . . is inaugural [moment] therein" of modern science, but also that "science continues to strengthen it [subject position] ever further" (Lacan 1965, 5).

Lacan's argument about the subject of science has two more contexts. Firstly, he argues that Descartes' *cogito ergo sum* is neither about knowledge nor about existence but about certainty. Lacan attributes Descartes' desire for certainty through the exercise of reason as significantly forming the modern subject. Secondly, Lacan further argues that this subject of modern science is split between knowledge and truth. For Descartes, while distinct observations are prerequisite for the construction of knowledge, it does not grant truth in itself; in other words, knowledge is not inherently true. Descartes invokes God – a non-deceiving agency outside of himself – to guarantee the truth of knowledge. The subject then pursues his own reason in pursuit of knowledge, and God guarantees the truth of this knowledge. Knowledge and truth are therefore separate; in fact, truth is external to knowledge, and only God joins them together. But God – the benevolent, good-natured God – is an assumption, in fact it is only the faith of the thinking subject (Lacan 1965, 35, 1977 [1964]).[6] It is the subject who imagines God who then grants truth to knowledge. The subject of science, Cogito, therefore is split between knowledge and truth and it is joined together by the (subjective) faith in God as symbolic (and not real) guarantee of truth outside of knowledge.

In discussing truth-to-nature epistemic objectivity followed in the seventeenth and eighteenth centuries, Daston and Galison also point out how it was God in whose image the truth or certainty of knowledge was granted. In the pursuit of perfect, pure, average, typical, and ideal images in nature, the scientist-subject was searching God's law that was worthy of admiration (Daston and Galison 2007, 68). The search for the eternal and non-deceiving agency to grant the truth signified the split between knowledge and its truth in the sense that truth was always external to knowledge. Simultaneous pursuit of knowledge and disavowal of its truth thus characterized the subject of science.

However, in modernity, the reference made to God to guarantee truth has significantly been altered. In fact, the God as truth-guarantor (in other words, God as symbolic guarantor) has acquired different names in the secular world. The faith in God is rearticulated and in turn placed in science itself. This faith is placed not just in science but in science's subjects, in the scientific community's confidence to deal with complex matters, in mechanical devices, in scientific method such as empiricism, logic, or mathematics. Alternatively, this faith to guarantee truth is entrusted to objects of science such as neurons, genes, or forces of nature (Glynos 2002, 65–66). These objects of science in fact acquire the place of demigod, for instance, the way the gene is made to comprise the immortal essence of humanity in Richard Dawkins' work on the selfish gene (Dawkins 2006).

In "Science and Truth", however, Lacan makes the point that the simultaneous pursuit of knowledge and the disavowal of its truth characterizes the subject of modern science. In making the methods and objects of knowledge to provide certainty and truth, modern science has progressively reduced truth to knowledge, in the sense that truth is increasingly located not outside but inside knowledge. And as a result, Lacan argues, the modern Cartesian subject split between knowledge and truth is erased or sutured. One of the main messages of "Science and Truth" is the way modern science effectively enables the scientist to forget his subjectivity, it actively forgets the subjective drama of its practitioners, "let's say that [science's] subject is not often studied" (Lacan 1965, 18). Lacan further argues that the individual consequences of forgetting subjectivity are manifested in mental anguish, suffering, even madness. Lacan counts Cantor (quoted at the beginning of the chapter) and Mayer in the list of such "first-rate tragedies" (Lacan 1965, 18).

Although Daston and Galison also arrive at the similar conclusion – in their argument, the demands of objectivity make the scientist-subject to deny his or her subjectivity, and many such scientists expelling their subjectivity in pursuit of representing nature as real became "well-nigh maniacal" (Daston and Galison 2007, 66) – there is a significant difference between theirs and Lacan's argument. For Daston and Galison, the erasure of subjectivity ensures objectivity, but for Lacanian psychoanalysis erasing or forgetting subjectivity means paranoia.

For Lacan, not just scientific but all knowledge is paranoiac. And still there is a significant difference between the paranoia caused by the mode of modern scientific practice and the foundational aphorism of "all knowledge being paranoiac" in Lacan's intellectual oeuvre. Paranoia is derived from Greek: para means outside or beside oneself and therefore beyond intelligible thought. For Freud, paranoia is "the blindness of the seeing eyes not wanting to know" (Freud 1893–1895, 117). In other words, what is paranoid is that which stands in opposition, alien to the self. It is not just in Lacan but also in the entire Freudian tradition of psychoanalysis, knowledge stands in dialectical relation to fear, the fear of not-knowledge. And, not-knowledge means negation, conflict or suffering that knowledge can cause. "My knowledge started off from paranoiac knowledge", says Lacan (1988 [1954], 163). In Lacan's work, the epistemological context of paranoiac knowledge is situated in three contexts or orders of being – imaginary, symbolic, and real. Firstly, knowledge is paranoiac developmentally. It begins with the "mirror stage" because it is acquired through our imaginary relation to the self and the other. In Lacanian psychoanalysis, the "mirror stage" that occurs between 6 and 18 months of infancy is the initial point of self-discovery when the nascent ego or "I" is discovered in the looking glass or through the eyes of another as metaphorical representation of the mirror. "The original specular foundation of the relation to the other . . . is the first alienation of desire" because it is fundamentally rooted in the imaginary. The ego first recognizes itself in an object outside of itself in the realm of the imaginary, in the illusory order, "in a fictional direction", and hence the source of alienation and paranoia (Lacan 1988 [1954], 176). Secondly, the Lacanian subject discovered in imaginary but lost in alienation is further recovered through the Other – the symbolic. "I" is never absolute or autonomous apart from the Other. The Other is "socially elaborated situations" mediated by linguistic structures, "I" is linked to the "Other", "I" exists as the Other (Lacan 1977 [1936], 5). However, the Other – to repeat, socially elaborated situations mediated through linguistic structures – especially the Other's desire is a potential threat to the subject, because it is an alien force, and hence the source of paranoia (Lacan 1988 [1954], 185). The third order of being, the "real", is the place of limit, the realm of unconscious, that which is lacking articulation in the imaginary and symbolic orders. The experience of knowing the real is the most horrific because it can never be known in itself. Freud compares the realm of the unconscious with the realm of the nature, "the unconscious is the true physical reality, in its innermost nature it is as much unknown to us as the reality of the external world" (Freud 1900, 613). In Lacan's work, therefore, all knowledge is paranoiac because the one that emerges from the orders of imaginary and symbolic is loaded with alienation, opposition, and demand. However, this knowledge is para-noiac knowledge not because of the fear of the unknown, but because it is tainted by the fear of knowing a particular truth of the self and the Other, this truth that the subject may find horrific. But in the domain of the real that the knowledge becomes impossible, because it acquires indescribable language, this knowledge is paranoiac because it is beyond mind, unknow-able, inaccessible. Because this chapter engages with the world of the real as represented in modern science, I have not given due space to the paranoiac knowledge of the subject emerging from the domains of the imaginary and symbolic and have focused more centrally on paranoia of the subject of sci-ence wanting to know the real.

History of objectivity in Daston and Galison's detailed work shows how in transiting from practices of truth-to-nature to mechanical objectivity all traces of God or *symbolic* guarantee is replaced with *real* guarantee. The shift from the truth-to-nature to mechanical objectivity is signified by representing nature not as ideal and perfect version that the God's law has created but as nature really is. Modern science developed increasingly sophisticated experiments and other methods to capture the reality of the real and to establish certainty of knowledge which almost always failed to guarantee such certainty and truth, and in this sense the real almost always remained unknowable. Daston and Galison discuss the way several scientists accounted for this impossibility of representing the real, some examples of which I have already mentioned. The other side, or perhaps the precondition of this search for the real, is what Lacan said, "there is no Other of the Other" (Lacan 1977 [1964], 77). In other words, there is no real guarantor behind the symbolic guarantor, first God and then science – God is only a matter of faith and so is science. It means how history of sci-ence since the time of Descartes has moved in the direction of foreclosing the symbolic guarantee for the truth of knowledge and replacing it with the real guarantee. As long as the scientific practice tries to find the real behind the symbolic, it will end up with "elaborate" and "grander" theories to bridge the gap, it will end up with proliferation of scientific knowledge and unending displacement of one scientific object for investigation to another. And in this sense the modern scientific practice resembles para-noia, not only because the knowledge of the real is beyond mind, but also because paradoxically this impossibility produces highly elaborate, rigor-ously logical, but delusional systems of knowledge (Glynos 2002, 65).

Relating this discussion to Daston and Galison's *Objectivity* would mean that the practice of objectivity emerges from the Cartesian subject's need for certainty. The paranoia that this need ultimately generates makes the "affective", the psychological, preceding and driving the epistemological.

Pathos, not ethos, drive epistemology

The scientist-self of mechanical objectivity that Daston and Galison por-tray matches nothing less than the basic principles of Kantian ethics – the categorical imperative (the duty qua wilful rejection of subjectivity) of the scientist-self is on the same side as the moral law (ethics of objectivity). In fact, the rational will of the self is the sole possible authority of the moral law – in other words, there is a complete match between the will and the moral law. Not only that the will of the scientist, which includes extreme modification of the self, is something inherently good, but that the moral law (the ethics of objectivity) is also unconditionally good in itself. In Kantian ethics, the categorical imperative is on the same side of the good (well-being) of the fellow-men. How ethics of objectivity will ensure the good of the people is not clear in *Objectivity*. The same question that is often posed against Kantian ethics can also be raised here: What if duty and good are on the opposite sides? What if duty could be accomplished at the detriment of the fellow-men? In other words, the notion of the good also heralds a pertinent question: Whose good? Daston and Galison, however, are clear on this point. They think that we must first know what objectivity is before we decide if it exists or it is good or bad. This is a fair argument.

To explore this logic further, in Kantian ethics, duty is only that the subject accepts as her duty, it does not exist in an external list of, say, Ten Commandments. The moral law is not some entity that prescribes "do this" or "do that" but it is the law that commands to perform duty without even naming it. In other words, the subject cannot claim that the duty was imposed upon her, that her action only followed the commandment of the law. The subject is responsible for her duty; she cannot hide behind the externally imposed moral law. Otherwise the ethical subject is just an agent of the moral law, an unnecessary and dispensable element of the moral law. Making the duty as one's own is at the core of the relation between the will and the moral law. To relate this reading to *Objectivity*, Daston and Galison's argument of the co-implication of objectivity-subjectivity perfectly matches with Kantian ethics in the sense that the ethical subject does not bring into a moral situation all subjective and affective baggage (which then needs to be removed or expelled) but the scientist-subject is born out of the moral situation – born out of the demands of the objectivity, she emerges from the moral situation of requirements of objectivity, she is made of, she is constitutive of the law of the objectivity. According to Kant, therefore, the subject acts contrary to her well-being and pathological interests only for the reasons of the moral law. In this sense the subject then performs an ethical act, i.e., when she goes against her well-being and pathologies, she is either angelic or diabolical – neither of which could be applied to human beings – and hence Kant says that an ethical act that goes against the well-being and pathologies of the subject is impossible (Zupančič 1998, 59). This impossibility pushes the subject to the manifestation of doubt and guilt and heralds a further indefinite struggle to separate herself from the pathology.

In the critical scholarship on Kantian ethics, it has been a long-standing question if Kant thought the reason qua will alone as the impetus for moral motivation or did he think there existed desire, feeling, or affect, prior to the will, as source of motivation to moral law. Is reason independently motivating or does it require a priori "affect" for the incentive for motivation? Kant himself wonders how it is possible for the "will without an object of representation" to be the direct incentive for the moral law and concludes that "this is an insoluble problem of the human reason" (Kant 1993, 75). There are scholars who insist that Kant put ethical principles as the absolute source of moral conduct (Nagel 1970, 11, 13), and there are others who think some other motivational factors beyond reason alone are necessary to account for moral obligation (Bond 1983, 11). In the chapter "Incentives of Pure Practical Reasons" in the *Critique of Practical Reason*, Kant meditates on "in what way the moral law becomes an incentive" for the will (Beck 1960, 217). But, here, contrary to his usual position on how moral law motivates human will through pure reason, Kant talks about the effects of the objective moral law on the human subject. This is because the human constitution is made of both a sensuous nature causing incentives and a rational nature which gives moral commands. And, in the human being, the commands of reason conflict with the inclination or incentives of sensuous nature. The freedom to act independently of sensuous inclinations results in pain – this, Kant called "negative subjective effects". These negative subjective effects are pathological in nature because they are derived from our sensuous nature. The negative subjective effects are not the only source for pathology, though (Beck 1960, 216). Kant also acknowledged that human actions are governed by another law, the law of the faculty of desire.

> Life is the faculty of a being by which it acts according to the laws of the faculty of desire. The faculty of desire is the faculty such a being has of causing, through its representation the reality of the objects of these representation.
>
> (Kant 1993, 10)

The faculty of desire is represented by a certain object, which could be shame, honour, fame, approval of others – on the contrary, the faculty of will is not represented by any such object. And the subject is "affected" by these representations, which is the cause of her action. For Kant, when affection is the cause of action, the actions are determined pathologically. Kant also talks about the "positive subjective effect" of the sensuous nature, which he calls "respect" for the moral law, or the moral feeling, which in German he calls *Achtung*. Kant attributes the correspondence of the moral law with the will to the presence of what he calls the "feeling of respect". The feeling of respect means that the law is nearby; it indicates the presence of the law; it provokes the feelings of fear, admiration, wonder, and awe – a feeling of sublimation. Although for Kant, the feeling of respect (for the moral law) "is a singular feeling, reason which cannot be compared with any pathological feeling", it is the only drive of *pure* practical reason (Kant 1993, 74–75).

This debate over the "pathological" component in Kantian ethics continues depending upon which part of Kant's vast array of intellectual work is invoked. For instance, Zupančič contends how Kant did not see that the feeling of respect could turn into pure and simple, *Ehrfurcht*, i.e., wonder – something that Kant himself linked to fear, thus becoming a perfectly pathological motive (Zupančič 1998, 67). Sytsma, on the other hand, discussing the role of *Achtung* in Kant's moral theory, interprets how "respect" is produced solely by reason, and thus respect for the law is not the incentive to morality but morality itself and hence far from pathological (Sytsma 1993, 121).

To sum up this discussion, firstly, for Kant a pure ethical act that goes against the well-being and pathologies of the subject is not possible; secondly, what Kant calls pathologies are objects of representation such as shame, honour, fame, and approval governed by the faculty of desire; and thirdly, the respect for the moral law is the only pure drive for practical reason.

In this context, many of Daston's *affect-saturated values* would qualify as pathologies in Kantian ethics. In *Objectivity*, Daston and Galison argue that the whole purpose of making atlas within the broad scope of truth-to-nature and mechanical objectivity was to establish standards for the entire disciplinary community for generations to come (Daston and Galison 2007, 202). Atlases were one among many scientific values and practices that meant to bind the dispersed members together and generate a sense of loyalty among them. "Internalized and moralized, these loyalties stamped a distinctly scientific self, which are recognizable across a diverse range of local contexts", say Daston and Galison (Daston and Galison 2007, 203). To repeat, in Kantian ethics the feeling of respect, the feeling for admiration and awe for the moral law, which in Daston would mean demands of objectivity, is the only pure drive of practical reason. Any other thing, even the affect of loyalty or desire for approval of others, would mean pathology.

As I have already discussed, responding to the reviewers, Daston and Galison argued that the notion of *affective states* are the ones in which the psychological and the epistemological, on the one hand, and the individual and the collective, on the other, are made to converge. Daston and Galison do not provide any explanation to the term *affective* neither do they make any gesture in relating it with the massive debates on the *affective turn* in social sciences, body studies, human geography, and media studies. The meaning of the affective is left to the imagination of the reader – and in my interpretation, it is broadly understood as the state that is embodied and deeply emotional but it is not reduced to the purely psychological.

Lorraine Daston in her work has written about what she calls *affect-saturated values* – a specific constellation of emotions and values, in a well-defined relation to each other, hanging in balance – upon which science depends for its practices (Daston 1995, 4). Similar to the arguments in *Objectivity*, the moral in the moral economy refers at once to the psychological and to the normative. Daston categorically emphasizes that moral economy is not a matter of individual psychology; she insists, "although the moral economies are about mental states, these are the mental states of collectives"; "it is not a matter of individual psychology"; "this is a psychology at the level of whole culture or subculture" (Daston 1995, 4). The examples of affect-saturated and shared values include devotion to clarity, accuracy, sociability among colleagues, impersonality, impartiality, moral obligation to discipline, diligence, fastidiousness, thoroughness, caution, integrity, trust, civility, curiosity, and honesty. Paul White comments that as the model of moral economy of affect-saturated values was refined in Daston's work, the affects retreated and morals held sway – "plenty of civility and very little feelings" (White 2009, 796). In the background of the discussion on Kantian ethics, all these collective affect-saturated values making epistemology would qualify as pathological rather than ethical, because in Kant all morality, by its definition, would exclude all pathological motives. To remind the readers, this pathology would include actions caused by the faculty of desire, represented by a certain object, for example, loyalty or fame or shame or approval. Although fame and shame are not listed in Daston's affect-saturated values, the point I intend to make is that even in Kantian ethics the pure act of reason, in other words, the moral law perfectly corresponding with the will is not possible. This would mean that even in Kantian logic the practices of objectivity are as much pathological as ethical. They are driven by pathos as much as by ethos.

Appearance is not being

Daston and Galison clarify that they are after understanding a certain "proto-typical knower" of nature, a type of scientist as a regulative ideal, as opposed to any flesh and blood individual (Daston and Galison 2007, 204). They are interested in delineating the normative force of historically specified exemplary persona (Daston and Galison 2007, 44). Metaphorically, they prefer "superficiality of enlargement" to "excavating conjectured depth". Also, their aim is to reveal patterns on how objectivity and subjectivity – two epistemologies, two ethics, two ways of life – intertwine so closely (Daston and Galison 2007, 205).

The enlarged view of the normative persona, however, does not necessarily coincide with the Foucauldian premise on *technologies of the self* that they adopt to explain the activation of such persona. In "The Moral Economy of Science", Daston postulates that the co-production of feeling and seeing happens in science schools by the way of apprenticeship – by making the members of the science inculcating self-discipline and self-surveillance on their own instead of being pushed or coerced. Daston refers here to Foucault's technologies of the self and clarifies that this is an exercise of a microscopic Foucauldian sort of power which structures the way the scientist comes to know (Daston 1995, 6). Daston and Galison emphasize the same point in *Objectivity* and argue that technologies of the self – the practices of mind and body – mold and maintain a particular kind of self (Daston and Galison 2007, 198–199). These practices include training the senses in scientific observation. Therefore, keeping lab notebooks and drawing specimens go hand in hand with monitoring one's own belief and quieting the will (Daston and Galison 2007, 199). As already mentioned, Daston and Galison clearly declare that they are interested in the normative force of regulative ideal and not in the flesh and blood individual. However, in arguing that the normative ideal moulds and shapes the self, and that it compels the scientist-self to quiet the will and monitor the emotions, Daston and Galison take *appearance for being*. They assume that the normative and regulative ideal is perfectly realized in the scientist-self being moulded and shaped in the processes of apprenticeship.

Based on Lacanian psychoanalysis' challenge to Foucauldian power and knowledge, I aim to argue in what follows that appearance and being never coincide. Instead, the discordant relation between appearance and being is not only the condition for desire but it explains the existence of conscience. And the place of the social surface – in terms of Daston and Galison's enlarged normative view – and the place of desire or conscience or guilt of the flesh and blood historical subject could be in contradiction.

Firstly, I aim to argue that under the panoptic gaze of technologies of the self the scientist-self becomes fully visible, governable, and tractable. In fact, the scientist-self becomes visible not only to others but to oneself through the specific, historically constructed norms and standards. However, what is implied in the argument that the scientist-subject is constructed by the panoptical gaze is that the subject thus is constructed by one monolithic discourse. The panoptic argument is ultimately resistant to resistance – in other words, it disallows a subject that can transcend the regime of power.[7] Also, the panoptic argument ignores the fact that the subject produced by the signifying system can never be determinate. The subject is indeterminable by its articulation that may result under the influence of a multitude of different discourses (Sangren 1995). Secondly, Joan Copjec in her Lacanian critique of Foucauldian historicism argues how the scientist-subject is formed in and by the field of science, but that the subject is never fully formed in this way. She refers to Bachelard to argue that the scientist-subject is split between two modes of thoughts, one governed by historically determined scientific norms and the other that are eternal, spontaneous, and purely mythical (Copjec 1994, 20–21). Copjec further discusses how Bachelard counted the obstacle of imaginary as the reason for only the partial success for the field of science to form the subject. Judith Butler elaborates on this point in her meditation on *The Psychic Life of Power* – it was Louis Althusser who further advanced Bachelard's theory to show that the category of imaginary is not an obstacle, on the contrary, it is absolutely necessary and an integral part of the historical process of construction of the subject (Butler 1997, 111–113). Daston and Galison do not account for the role of the imaginary in the historically determined ideological formation of the subject and therefore their argument depends upon the assumed verisimilitude between the ideal, the regulative, the normative, and the real referent. Although they have emphatically declared that their inquiry is focused on the normative scientist-subject formed by the field of objectivity and

not on the actual social individual, the Foucauldian argument of technologies of the self could be understood only with reference to the actual
social setting, however hypothetical it might be. In fact, in Althusser, the
imaginary, a correction on Bachelard, almost exclusively bears the burden
of ideological construction of the subject. Meaning, the imaginary is constitutive of the ideological. On the contrary, the categories of imaginary
are barely given any space in Foucault's, and by their following Daston
and Galison's definition of the normative laws that govern technologies
of the self (Copjec 1994, 20–24). The law is thus unconditional, it must
be obeyed; in fact, *being* by definition is obedience. Hence, although the
images of self-surveillance and self-correction are required to construct
the subject, if the subject thus constructed is absolutely upright and correct, fully formed, these categories of self-surveillance and self-correction
should become redundant. The Foucauldian technologies of the self fail
to take into consideration the dialectics between the norm and desire and
the element of indeterminacy that desire introduces in the making of the
subject. And therefore, the enlarged normative view would have made
proper sense only if it were juxtaposed with the deeper excavation of how
these norms actually work out in the individual scientist's life.

The role of unconscious in the making of 'science as habit'

In locating the epistemic fear as an affective state, the book *Objectivity* makes
a connection with Daston's earlier works on "The Moral Economy of Science" (Daston 1995) and with her later work "On Scientific Observation"
(Daston 2008). In the latter, Daston reflects on psychology and proposes
to bridge the gap between psychology and epistemology – clarifying again
that psychology in question was collective rather than individual (Daston
2008, 97). She argues here that learning to see like a scientist is a matter of
accumulated experience, a matter of habit formation of a well-trained collective. But acquiring this experience is an equation of time – the scientist
has to go through a gradual process of training before experience turns into
a habit whence for a mature scientist it becomes possible to see things all at
once – a naturalist in 1922 described it as a "jizz", informs Daston. The jizz
is an observation that is sure, swift, silent, and happens without a pause for
mental analysis (Daston 2008, 101). But before the scientist comes to the
position to be able to see things all at once, in a jizz, as a matter of habit, she
goes through a process of learning in which perception turns into memory
and experience. And, Daston clarifies, even when this is all about the production of conscious reason, "the faculty of epistemology has no inkling as
to how it's done" (Daston 2008, 102).

Referring to Ludwik Fleck's *Genesis and Development of a Scientific Fact*, Daston further emphasizes the trope of temporality and argues that this knowledge is not tacit, retroactively it is perfectly possible to describe this process of habit formation – the way perceptions coalesce into experience – in considerable detail, in a series of stages, and thus bring it to the modes of representation. However, while in action – while the scientist is seeing things all at once, in a jizz – the conscious reason and faculty of epistemology has no idea how this is done. Daston here refers to Ludwik Fleck's interpretation of thought collective and also genesis of the scientific fact. Fleck describes the process of genesis of the fact – it arises after "a signal of resistance in the chaotic initial thinking, then a definite thought constraint, and finally a form [of the fact] to be directly perceived" (Fleck 1981 [1935], 95). The genesis/crystallization of the fact in a thought collective is described in a musical metaphor – "the confused notes followed by hummed and inaudible tunes gradually turning into a melody once the 'co-workers' listened and tuned their sets until these became selective" (Fleck 1981 [1935], 95). The melody could then be heard even by the unbiased person – meaning, it crystallizes into a fact. The genesis of the scientific fact thus is not only a process in the sense that it is emergent through time, but most importantly, it involves a "chain of experiences" – first chaos, then constraint, then perception, and then fact emerges. There is similarity between Daston's experienced scientist able to see in a jizz, all at once, and Fleck's genesis of the scientific fact – both are processes, are emergent, subjected to time, and both constitute a chain of experiences that is although silent and swift in Daston, but can be retroactively traced into a series of stages in both Daston and Fleck.

However, there are subtle ways in which the irrational and rational, reasonable and unreasonable, conscious and unconscious, feelings and thought, experience and fact are intertwined in these accounts. In *Objectivity*, Daston and Galison refuse to call "epistemic fear" anything than epistemic, because it could otherwise open the door for murky realms of the "irrational"; the historians of science are very wary of these – Daston informs us (Daston 2008, 101). Here, the individual subjectivity, a specific emotional trajectory, the psychology of innermost thoughts, and other affective intentions of historical actors are relegated to mind reading, and to biographies of the individual scientists. For Daston, the epistemic fear is structural and collective and hence "there is nothing irrational (or reductive) about [it]", in fact, "it is simultaneously reasonable, psychological and collective" (Daston and Galison 2008, 672). In my reading, it is Daston and Galison's own fear of "irrationality" that makes them fear the epistemic fear for being anything else than epistemic.

But then, even after arguing to make the epistemic fear "not irrational", Daston still has to enter in the murky ground in which conscious
and unconscious and thought and feeling are inseparably intertwined – for
instance, the jizz happens without a pose for a mental analysis, here the
emergence of the conscious scientific reason has to go through the path
of the unconscious; the way perception turns into memory into experience into habit has a psychology. And yet again, the conscious reason,
the faculty of epistemology, has no inkling how it is done when it is done
(to repeat, the stages of this process can be retroactively traced, though).
This all-at-once-ness in Daston is explained in the words of Descartes, "the
arguments so speeded up that it bursts upon the mind as a single cognitive
event" and "[n]o amount of explicit reasoning, even mathematical reasoning, can compete with it" (Daston 2008, 110).

In these arguments of Daston the meaning of the psychological finally
arrives home. The jizz, all-at-once-ness, is not about some *random* erup-
tion of unconscious upon the conscious. This is about the unconscious
fundamentally structuring the conscious reasoning without the conscious
even knowing it. In fact, this is about the *indispensability* of the unconscious in ordering the conscious. The claims of all-at-once-ness, the jizz,
the arguments bursting upon the mind as a single cognitive event, "they
concern", Daston writes, "largely unconscious processes of perception"
happening "albeit processes that are consciously . . . taught and controlled
by the exercise of scientific observation" (Daston 2008, 105). And still,
instead of accepting this crucial role of the unconscious in structuring the
conscious, Daston's tone turns defensive, she insists that even when the
perceptual habit – the jizz and all-at-once-ness – "is not *of* reason", or more
than reason, "this does not render such habits *ipso facto* irrational" (Daston 2008, 105). She repeatedly asserts that, "[t]here is nothing individual,
nothing arbitrary, nothing mystical, [nothing irrational] about this kind of
psychology" (Daston 2008, 106). It's the fear of epistemic fear lapsing into
individual and hence subjective and hence irrational that drives Daston
and Galison to claim that the fear driving epistemology is purely and only
epistemic. I can't but mention here that Fleck, whom Daston significantly
refers to make her point, however, is upfront in accepting the "irrationality" of "being experienced". He says, "[t]he ability directly to perceive . . .
is acquired only after much experience. . . . The concept of being experienced, with its hidden *irrationality*, acquires fundamental epistemological importance" (italics mine) (Fleck 1981 [1935], 92). My claim is that
Daston's conceptualization of the terms like unconscious, psychological,
individual, and collective are underdetermined because incorporating their
meaning in all richness will threaten the master-self of Daston and Galison

that is made of the Kantian "ethical imperative" at the core. As already discussed at length, Daston and Galison insist that "the link between epistemology and ethics" is very strong. Whatever the epistemic virtue, "the exhortations are nearly always religious and ascetic in tone. There is always an ethical imperative at core" (Daston and Galison 2007, 40).

The role of the unconscious is truncated in Daston's work in the degree to what is needed to aid what Fleck calls the "chain of experiences" coalescing into "the jizz", and "the all-at-once-ness". In other words, the unconscious is there only to silently and swiftly structure the experienced scientist's epistemic conscious. Daston here employs the concept of the unconscious in a selective manner without referring to its proper origin either in the tradition of Freudian psychoanalysis or even prior to that tradition. The incorporation of the meaning of the unconscious as it is understood in the Freudian tradition would seriously jeopardize the self-mastery at the centre of the practices of objectivity in Daston and Galison's *Objectivity*. The Freudian subject is fundamentally split between the conscious and unconscious. In fact, the unconscious does not come to rescue, aid, or structure the agency of the conscious, but its repressed drives put serious obstacles in the path of the rational conscious, and these drives are obstinately resistant to control and even comprehension by the conscious. The agency of the unconscious, as it is projected in Daston, is something akin to "psychological unconscious" and not the genuinely Freudian unconscious, which Felicity Callard, discussing such debates in human geography, argues constructs an idealized psyche because it is far easier to deal with it and to make it susceptible to change and transformation, it is an accessible psyche, conducive to agency, and hence available for political manoeuvring. But this is not the way the psyche necessarily operates if the psychoanalytical concept of unconscious is genuinely taken into consideration (Callard 2003, 308). And if the Freudian notion of the unconscious is taken into consideration, the Kantian notion of self-mastery at the core of objective knowledge production will come under serious challenge.

Conclusion

The main focus of this chapter is to critique the neo-Kantian scientist-self in Lorraine Daston and Peter Galison's book *Objectivity*. While doing so, I have also closely engaged with Daston's later and earlier works "On Scientific Observation" and "The Moral Economy of Science", respectively.

The chapter challenges Lorraine Daston and Peter Galison's formulation of the neo-Kantian self at the core of the scientific-subjectivity. They posit that the scientist-self is first and foremost normative, it is determined to do the right thing, i.e., fulfil the ethos of objectivity. In other words, for Daston and Galison the "categorical imperative" (the duty qua wilful rejection of subjectivity) of the scientist-self is on the same side as the moral law (ethics of objectivity). Based on this conception of the master-self, Daston and Galison argue that the link between epistemology and ethics is so strong that whatever the epistemic virtue, there is always ethical imperative at the core. This scientist-self is affective (fearful) only in the noble pursuit of objectivity. In fact, the fear of subjectivity is the fear that the core self would become an obstacle in the demands of objectivity and hence this fear is foremost ethical and normative. Daston and Galison's scientist-self is thus made of active self-consciousness upon which both the ethics of knowledge and knowledge are built.

I have presented four challenges to Daston and Galison's central argument that epistemology and ethos fuse. Firstly, following Jacques Lacan's work, I have argued that the subject of science constituted by the mode of modern science suffers from paranoia. It is not the fear of subjectivity interfering with objectivity but the impossibility of knowing the truth of the real (of the nature) that causes paranoia. This paranoia, on the one hand, caused by the erasure of subjectivity triggers much suffering in the scientist-self, and, on the other, creates grander and elaborate scientific theories that almost always fail to provide the truth or certainty of knowledge. Here, I have argued that it is not ethos of objectivity that drives epistemology but pathos of paranoia. The second challenge builds upon Kant's own denial that the perfect correspondence between the human will and the moral law is possible. Kant himself thought that an ethical human act is impossible without the component of pathology. Thirdly, referring to the contestation of historicism in Foucault's work from the location of the existence of desire as the formative part of subjectivity, I posit that Daston and Galison take appearance for being. Daston and Galison have argued in *Objectivity* that they are concerned only with the prototypical knower, what they call the enlarged view of the normative persona, and not with the experiences of flesh and blood individuals in real historical and cultural settings. They, thereby, assume verisimilitude between the ideal, the regulative, and the normative and the real referent. In brief, the problematic of the indeterminate and often contradictory relationship between the normative and the real is pointed out. The fourth challenge questions the notion of psychological and unconscious in the making of epistemology in Daston's earlier work on "The Moral Economy of Science" and "On Scientific Observation", which also make close connection with Ludwik Fleck's *Genesis and Development of a Scientific Fact*. These challenges question, on the one hand, the neo-Kantian conception of the scientist-subject, and, on the other, contest Daston and Galison's argument that ethos and not pathos drives the imperative of epistemology.

Based on these challenges to Daston and Galison's formulation of the co-construction of psychology/subjectivity and epistemology/objectivity, I wish to claim, following Michael Polanyi (following Sartre), that *existence precedes essence*, or, in other words, psychology precedes epistemology. This chapter aims to make a case for the argument that understanding and disclosing "entities" in scientific domain presupposes an understanding of "being" in general; it also aims to make a case for an alternative conception of the scientist-subject and correspondingly an affective and existential conception of science that I have further explored elsewhere (Shah 2013, 2016).

Notes

1 This chapter was first published as Shah, E. 2017. "Who Is the Scientist-Subject? A Critique of the Neo-Kantian Scientist-Subject in Lorraine Daston and Peter Galison's *Objectivity.*" *Minerva*, 55 (1), 117–138 and is republished here with the journal's permission. https://doi.org/10.1007/s11024-017-9313-5.

2 As quoted in Burgoyne (2002: 237).

3 Hermann similarly proposed parallels between mathematical and psychic structures in the work of Bertrand Russell (Burgoyne 2002, endnote 25, 255).

4 Finding out *how we live as we do now* Steven Shapin explores the relationship between the authority of knowledge and the character of knowers (Shapin 1988, 1991, 2007). More so, the co-dependence of the cognitive and the emotional is also approached by the growing field of existentially oriented scientific biographies. Here, science is projected as an existential choice made by scientists (Nye 2006). See also Söderqvist (1996, 69). Lastly, the oldest of the inquiries on science and the self is made by the feminists. Do women scientists practice science differently than men? Would science be less reductionist, more empathic and intuitive if women had an equal role in shaping it? For a long time, feminist epistemologies have been engaged with theorizing the relationship between objectivity and subjectivity, in which the idea of what can be counted as objective science and who is the knowing subject have been revised from their positivist and foundationalist predecessors. See, for example, Harding (1986, 1987, 1994, 1996). Also see Longino (1993, 2001). For a detailed discussion on feminist epistemologies and their construction of the scientist-subject, see Shah (2013).

5 Dany Nobus attempts to "reconstruct and clarify Lacan's argument" in the text Science and Truth. Nobus clarifies that the expression of subject of science in this text is extremely ambiguous. It simultaneously refers to "the scientist, the topic of study within scientific practice, science itself, the subjective elements within science, and the objects subjugated to scientific investigation." I also find Science and Truth often beyond comprehension. I thereby have closely consulted Nobus' interpretation that subject of science juxtaposed with Cogito in the text means human rationality, mental power, and the certainty of continuous experience of thought (Nobus 2002, 94).

6 Lacan's Science and Truth, the transcript of the opening seminar on *The Object of Psychoanalysis*, was first published in 1965 (Lacan 1965). Many ideas in Science and Truth emerged in Lacan's previous seminar on *The Four Fundamental Concepts of Psychoanalysis*, which were subsequently summarized in the paper Position of the Unconscious. In this trail of thoughts emerging over a year, Lacan was concerned with the subject of science and with the scientificity of psychoanalysis but also about the way unconscious was structured like language. Many scholars, including those who are expert on Lacan's oeuvre have complained that the transcribed and translated text of Science and Truth is fairly unintelligible and inaccessible. It is crowded with a multiplicity of ideas presented in highly fragmentary ways. Lacan in general is a difficult philosopher to follow. For example, Jon Mills, one of the prolific commentators of Lacan's work, describes his writing as fragmentary, chaotic, incoherent, his jargon highly esoteric which resists articulate systematization (Mills 2003). Most of Lacan's writings were originally not written; they were transcribed from lectures and seminars which adds an additional layer of inaccessibility, not to even mention the problems created by the translation from French to English. One translator and a long-time critical scholar of Lacan's work, Bruce Fink, put it in the preface of his translation, "Lacan's French is . . . so polyvalent and ambiguous that some frame must be imposed to make sense of it whatsoever" (Miller 1975). Science and Truth has made sense to me in consultation with other interpretative contributions on "reconstruction and clarification of Lacan's argument" (Nobus 2002, 89). I have therefore referred to a few such efforts at "reconstruction and clarification" of Lacan's arguments in Science and Truth (Burgoyne 2002; Glynos 2002; Miller 2002; Nobus 2002; Stavrakakis 1999).

7 For the further discussion on resistance in Foucault's notion of power, see Sangren (1995).

References

Anderson, Amanda. 2008. "Epistemological Liberalism: Objectivity by Lorraine Daston and Peter Galison." *Victorian Studies* 50 (4):658–665.

Beck, Lewis White. 1960. *A Commentary on Kant's Critique of Practical Reason.* Chicago: University of Chicago Press.

Bond, E. J. 1983. *Reason and Value.* Cambridge: Cambridge University Press.

Burgoyne, Bernard. 2002. "What Causes Structure to Find a Place in Love?" In *Lacan and Science*, edited by Jason Glynos and Yannis Stavrakakis, 231–263. London: Karnac.

Butler, Judith. 1997. *The Psychic Life of Power: Theories in Subjection.* Stanford: Stanford University Press.

Callard, Felicity. 2003. "The Taming of Psychoanalysis in Geography." *Social and Cultural Geography* 4 (3):295–312.

Copjec, Joan. 1994. *Read My Desire: Lacan against the Historicists.* Cambridge, MA: The MIT Press.

Daston, Lorraine. 1995. "The Moral Economy of Science." *Osiris* 10 (2nd Series):2–24.

Daston, Lorraine. 2008. "On Scientific Observation." *Isis* 99 (1):97–110.

Daston, Lorraine, and Peter Galison. 2007. *Objectivity.* Brooklyn: Zone Books.

Daston, Lorraine, and Peter Galison. 2008. "Objectivity and Its Critics." *Victorian Studies* 50 (4):666–677.

Dawkins, Richard. 2006. *The Selfish Gene*. New York: Oxford University Press.

Fleck, Ludwik. 1981 [1935]. *Genesis and Development of a Scientific Fact*. Chicago: University of Chicago Press.

Freud, Sigmund. 1893–1895. *Studies on Hysteria*. London: Hogarth Press.

Freud, Sigmund. 1900. "The Interpretation of Dreams." In *Standard Edition*, vols. 4–5. London: Hogarth Press.

Glynos, Jason. 2002. "Psychoanalysis Operates upon the Subject of Science: Lacan between Science and Ethics." In *Lacan and Science*, edited by Jason Glynos and Yannis Stavrakakis, 51–88. London: Karnac.

Harding, Sandra. 1986. *The Science Question in Feminism*. New York: Cornell University Press.

Harding, Sandra. 1987. "Introduction: Is There a Feminist Method?" In *Feminism and Methodology*, edited by Sandra Harding, 1–15. Bloomington: Indiana University Press.

Harding, Sandra. 1994. "Is Science Multicultural? Challenges, Resources, Opportunities and Uncertainties." *Configurations* 2 (2):301–352.

Harding, Sandra. 1996. *Whose Science? Whose Knowledge? Thinking from Women's Lives*. New York: Cornell University Press.

Kant, Immanuel. 1993. *Critique of Practical Reason*. Translated by Lewis White Beck. New York: Macmillan.

Keller, Evelyn Fox. 1985. *Reflections on Gender and Science*. New Haven: Yale University Press.

Lacan, Jacques. 1965. "Science and Truth." *Newsletter of the Freudian Field* 3 (1/2):1–29.

Lacan, Jacques. 1977 [1936]. "The Mirror Stage as Formative of the Function of the I." In *Ecrits: A Selection*, translated by Alan Sheridan, 1–6. New York: Norton.

Lacan, Jacques. 1977 [1964]. "The Subject and the Other: Alienation." In *Seminar, Book XI: The Four Fundamental Concepts of Psychoanalysis*, edited by Jacques-Alain Miller, 203–215. New York: W.W. Norton & Company.

Lacan, Jacques. 1988 [1954]. "The See-Saw of Desire." In *The Seminar of Jacques Lacan, Book I: Freud's Papers on Technique, 1953–1954*, edited by Jacques-Alain Miller, 163–175. Cambridge: Cambridge University Press.

Lacan, Jacques. 1995 [1964]. "Position of the Unconscious." In *Reading Seminar XI: Lacan's Four Fundamental Concepts of Psychoanalysis*, edited by R. Feldenstein, B. Fink and M. Jaanus, 259–282. Albany, NY: State University of New York Press.

Longino, Helen E. 1993. "Subjects, Power, and Knowledge: Description and Prescription in Feminist Philosophies of Science." In *Feminist Epistemologies*, edited by Linda Alcoff and Elizabeth Potter, 101–120. London: Routledge.

Longino, Helen E. 2001. "Can There Be a Feminist Science?" In *Women, Science, and Technology: A Reader in Feminist Science Studies*, edited by Mary Wyer, Mary Barbercheck, Donna Giesman, Hatice Örün Öztürk and Marta Wayne, 207–213. New York: Routledge.

Miller, Jacques-Alain, ed. 1975. *The Seminar of Jacques Lacan: On Feminine Sexuality, the Limits of Love and Knowledge, Book XX Encore 1972–1973*. New York: W.W. Norton & Company.

Miller, Jacques-Alain. 2002. "Elements of Epistemology." In *Lacan and Science*, edited by Jason Glyson and Yannis Stavrakakis, 147–166. London: Karnac.

Mills, Jon. 2003. "Lacan on Paranoiac Knowledge." *Psychoanalytic Psychology* 20 (1):30.

Nagel, Thomas. 1970. *The Possibility of Altruism*. Princeton: Princeton University Press.

Nobus, Dany. 2002. "A Matter of Cause: Reflections on Lacan's 'Science and Truth.'" In *Lacan and Science*, edited by Jason Glyson and Yannis Stavrakakis, 89–118. London: Karnac.

Nye, Mary Jo. 2006. "Scientific Biography: History of Science by Another Means?" *Isis* 97 (2):322–329.

Sangren, Steven. 1995. "'Power' against Ideology: A Critique of Foucauldian Usage." *Cultural Anthropology* 10 (1):3–40.

Shah, Esha. 2013. "Rosalind Franklin and Her Science-in-the-Making: A Situated, Sexual and Existential Portrait." *Yearbook of Women's History/Jaarboek voor vrouwengeschiedenis* 33:127–146.

Shah, Esha. 2016. "A Tale of Two Biographies: The Myth and Truth of Barbara McClintock." *History and Philosophy of the Life Sciences* 38 (18):1–12.

Shapin, Steven. 1988. "The House of Experiment in Seventeenth Century England." *Isis* 79:373–404.

Shapin, Steven. 1991. "'A Scholar and a Gentleman': The Problematic Identity of the Scientific Practitioner in Early Modern England." *History of Science* 29:279–327.

Shapin, Steven. 1992. "Discipline and Bounding: The History and Sociology of Science as Seen through the Externalism-Internalism Debate." *History of Science* 30:333–369.

Shapin, Steven. 2007. *The Scientific Life: A Moral History of a Late Modern Vocation*. Chicago: University of Chicago Press.

Shapin, Steven. 2012. "The Sciences of Subjectivity." *Social Studies of Science* 42 (2):170–184.

Söderqvist, Thomas. 1996. "Existential Projects and Existential Choice in Science: Science Biography as an Edifying Genre." In *Telling Lives in Science: Essays in Scientific Biography*, edited by Michael Shortland and Richard Yeo, 45–84. New York: Cambridge University Press.

Stavrakakis, Yannis. 1999. *Lacan and the Political*. London: Routledge.

Switzer, Adrian. 2009. "Review of *Objectivity*." *Foucault Studies* 6:96–104.

Sytsma, Sharon. 1993. "The Role of Achtung in Kant's Moral Theory." *Auslegung* 19 (2):117–122.

White, Paul. 2009. "Introduction, Special Issue on Focus: The Emotional Economy of Science." *Isis* 100 (4):792–797.

Zupančič, Alenka. 1998. "The Subject of the Law." In *Cogito and the Unconscious*, edited by Slavoj Žižek. Durham and London: Duke University Press.

3

IMMORTALITY IDEOLOGIES AND THE PARTICULATE GENE

H. J. Muller

I had made my way through a dark maze, but it was the bright City of the Immortals that terrified and repelled me. A maze is a house built purposely to confuse men; its architecture, prodigal in symmetries, is made to serve the purpose. In the palace that I imperfectly explored, the architecture had no purpose. There were corridors that led nowhere, unreachably high windows, grandly dramatic doors that opened onto monk like cells or empty shafts, incredible upside-down staircases with upside-down treads and balustrades. Other staircases, clinging airily to the side of a monumental wall, petered out after two or three landings, in the high gloom of the cupolas, arriving nowhere. *This City, I thought, is so horrific that its mere existence, the mere fact of its having endured – even in the middle of a secret dessert – pollutes the past and the future and somehow compromises the stars. So long as the City endures, no one in the world can ever be happy or courageous.*

(Borges 1998)

This paragraph is quoted from the short story *City of the Immortals* by Jorge Luis Borges. Borges' story is an autobiographical tale of a Roman soldier Flaminius Rufus who arrives in the city of the immortals and discovers that the immortals, a tribe of Troglodytes, have no language, no culture, no pity; they live in sand caverns like quasi-animals. The quoted paragraph depicts how the lack of purpose in the city of immortals translates into the absence of order and pattern; how the experiences of irregularities, endlessness, chaos, heterogeneity – in other words, irrationalities – turn into experience so horrific and oppressive for Flaminius Rufus, who is a mortal turned immortal, that he runs away from this horror and wishes to retain no memory of seeing the city of immortals. Borges in fact connects purpose and rationality with knowledge of mortality – the intolerable city of immortals is founded on the lack of knowledge of human finiteness.

Borges' city of the immortals lacks order and pattern, while, in contra-distinction, the world of science-based civilization of the mortals is insti-tuted on the metaphysical idea of a profoundly orderly universe. Following this contradistinction, the starting point of this chapter is a question: How could the origin of these metaphysical presuppositions of the universe being a determinate and orderly place be explained? My central claim in this book is that a scientist's particular ways of being in the world, including her metaphysics, pioneer the structures of rational and cognitive thought. My claim is that scientists look for conceptual worlds that respond to their "emotional longing", or in other words, intellectual paradigms are *affect worlds*, the conceptual theories are *isomorphic* with the world emotion-ally and existentially desired.[1] Accordingly, my aim in this chapter is to show how the founding moment of the concept of the gene in the work of pioneering scientist Hermann Muller was rooted in the existential idea of human immortality. I engage with Muller's life and thought diachronic-ally in the context of his life-project and demonstrate how the immortal-ity ideologies, although not so consciously and rationally but nevertheless, were *a priori* informing and forming his theoretical and empirical molecular biology. The biographical portrait of Muller is projected in the background of the collective projection of science for humanity during the interwar period.

"Being" and "death": immortality ideologies

The awareness of death is an integral part of the existential human being. "Humans are the only creatures who not only know, but also know that they know – and cannot 'unknow' their knowledge of mortality" says Zyg-munt Bauman. Bauman further argues that awareness of mortality is the ultimate condition of cultural creativity which creates conditions for per-manence, the way to defy the finiteness and mortality (Bauman 1992, 4–5). Immortality, that is, denial and defiance of death – have played a major role in Western philosophy at least since the seventeenth century. Here I only briefly review some of those pioneering influences that are relevant for the history of science explored in this chapter.

The fear of death has played a pioneering role in the civilization-making. Thomas Hobbes' political philosophy of the modern state was based on the theory that the desire for security – in other words, overcoming the fear of (violent) death – is the most fundamental and rational human desire and the modern state based on satisfying this desire will keep the anar-chy at abeyance. The political theory of the modern state, the Leviathan, now come to be known as the "social contract theory" can be described as the method by which rational, equal, and free individuals come to agreements about the political principles and arrangements for coexistence. Hobbes' theories on the fear of death and the modern state were crucially influenced by a classical thinker Thucydides whom Hobbes greatly admired. There is, however, a fundamental difference between Hobbes' and Thucydides' theories on human nature. Thucydides believed that the hope for immortality – the hope that one can overcome mortality and live on after death – overwhelms human nature and not the fear of death. The hope for immortality, as Thucydides argues, takes either the form of living on through a formation of city or nation or living on in the memory of others by winning the reward of eternal fame. Thucydides believed that not only the hope of immortality is stronger than the fear of violent death but that the threat of death does not dampen but inflame the hope for immortality. In other words, the hope for immortality could be far more powerful than the fear of death and it can undermine the elaborate system of the Hobbesian modern state founded on the fear of death.[2]

Otto Rank, a psychoanalyst who was a contemporary and dissident of Freud, instead associates the awareness of death with the irrational self. He ascribes the profound awareness that the world and events in one's own personal life are finite, inexplicable, and uncontrollable to the irrational self. This irrational self finds expression in "seeking material immortalization in lasting achievements" (Rank 1958, 84). And despite this awareness of finiteness, the efforts of mastery, for example of scientific knowledge, are overwhelmingly governed by the rational at the expense of the irrational.[3]

In comparison, Heidegger's existential ontology of Dasein (being in the world)[4] is structured around incorporation of the lack and anxiety that the awareness of death generates. Dasein is being-toward-death in the sense that death is the basic certainty of Dasein – "the MORIBUNDUS first gives the SUM its essence".[5] In other words, the constancy and all-pervasiveness (awareness) of death individualizes the human self. In fact, in Heidegger, the authenticity of Dasein requires acceptance of one's own death, because Dasein's totality can be revealed only in its being-toward-death.[6] Dasein in fact existentially experiences total powerlessness, vulnerability, and anxiety on the face of the ultimate, inescapable, threat of death. In response to all-pervasiveness of death, "care is the basic state of Dasein".[7] Dasein is care because Dasein is always concerned about its being on the face of the indefiniteness of the moment of death which leads Dasein to look for the unshakable certainty; it leads Dasein to do everything to deny its mortality; holding death as ultimate truth permeates all of its attitudes and stances. In fact, the anxiety-ridden awareness of being-toward-death gives the Dasein its temporality; this awareness makes the Dasein always being ahead of itself, future-oriented, projected towards a field of its possibilities; Dasein is always awaiting and expecting, hopefully and fearfully. Dasein's understanding of the world is projecting into possibilities.

Based on this conception of Dasein, Heidegger clarifies the existential conception of science. In contrast to the idea of logical science that amounts to science as empirical results and outcomes, the existential conception of science emphasizes scientific possibilities. In this sense science is not accumulation of verifiable knowledge but it is always directed ahead in the future towards possibilities it cannot yet fully grasp. Integral to futuristic orientation of science and knowledge thus is Dasein's way of being-toward-death. Scientific understanding is therefore integral to a larger historical disclosure of possibilities. And scientific practices transform their issues and stakes corresponding to such possibilities. Because scientific understanding, be it focused on specific phenomena, experimental system, or theoretical concept, is always oriented towards a subsequent advance.[8]

Heidegger's existential ontology of authentic Dasein, who has the courage to face and not fear the death, however, is impossibility according to Otto Rank's empirical anthropology of immortality ideology. Rank extensively studied ancient and modern human history and argued that "history can be understood as the succession of ideologies that console for death", and that no human society is able to overcome the fear of death and in fact all human societies alleviate the fear of death through what he called soul beliefs.[9] A soul belief is a belief anchored in individual, collective, or generative immortality, for example, eternal life granted by the doctrine of Christian salvation or through procreative acts by begetting children. Rank is most provocative when he says that the soul-belief is replaced in modern times by scientific intellectualism – the new god of truth. Rank thinks that "all truth-seeking is in the end the old struggle for the soul's existence and its immortality".[10] Ernest Becker elaborates on Rank's observation in two of his well-known publications, *The Denial of Death* and *Escape from Evil*, and argues that

> what people want in any epoch is a way of transcending their physical fate, they want to guarantee some kind of indefinite duration and culture provides them with the necessary immortality symbols or ideologies; societies can be seen as structures of immortality power.
>
> (Becker 1975, 63)

Zygmunt Bauman makes a similar point. He propounds a hypothesis that "social institutions and cultural solutions are sediments of the processes set in motion by fact of human mortality" and he pleads to open up an investigation on the presence of death (the conscious or repressed knowledge of mortality) in human institutions, rituals, and beliefs so that the fact of mortality and the knowledge of it can be dealt with and processed and turned into a major source of life's meaning. For Bauman, the "life forgetful of death . . . and alive with purposes instead of being crushed and incapacitated by purposeless" is a formidable human achievement (Bauman 1992, 4–10).

In sum, following inferences are derived from the preceding discussion on the importance of immortality ideologies for the philosophy of science: 1) science as a social and cultural institution is a process set in motion by the fact of human mortality; in modern times science is the principal culture that provides symbols and tools to transcend the physical fate, to guarantee immortality; 2) all truth-seeking, scientific intellectualism being the latest form of truth-seeking in the history of human civilization, is in the end the old struggle for the soul's existence and its immortality; 3) the fear of the irrational, the profound awareness that the life and world are finite, inexplicable, and uncontrollable governs the production of rational knowledge; 4) corresponding to the existential condition of Dasein, the being-toward-death, the empirical science ultimately feeds into existential conception of science, i.e., science is always directed ahead of itself, it is future oriented, projected towards field of possibilities; and, finally, 5) the scientist nourishes her immortality in self-perpetuation; she is even willing to die perpetuating her science as ideology.

Eternity of the atom and the gene and human immortality

At the founding moment of the concept of the gene, the mood was exuberant in terms of the "future possibilities" the nature and structure of the substance of heredity (the gene) was unravelling. For instance, J. B. S. Haldane, a leading population geneticist, also a member of the British Socialist Party, in his 1923 popular article titled *Daedalus: Or, Science and the Future,* speculated the influence of such biology on history in 150 years.[11] Haldane writes here like a historian writing a century and half later,

> As early as the first decade of the twentieth century we find a conscious attempt at the application of biology to politics in the so-called eugenic movement. A number of earnest persons, having discovered the existence of biology, attempted to apply it in its then very crude condition to the production of a race of supermen. They appear to have managed to prevent the transmission of a good deal of syphilis, insanity, and the like.
>
> (Haldane 1923, 14)

Haldane refers to Thomas Morgan's research on *Drosophila* (a common fruit fly) chromosome – which was in its very infancy at the time the article was written – to extrapolate it to the production of a race of super-humans. He speculates the future history supposed to have been read by undergraduate students in 150 years,

> how little the bearing of the results such as Morgan's work on Mendelian factors in the nucleus of Drosophila in 1912 was anticipated to have such remarkable influence on the biology of future. In 1951 the first ectogenetic child was produced.
>
> (Haldane 1923, 15)

The technique of ectogenesis was imagined as a method to create a child in artificial fluid. The child was born out of eggs produced from an ovary also preserved in such fluid. Haldane goes on to explain in his futuristic history the way 60,000 such children were annually produced in France by 1968. Further, the history 150 years in the future continues, "as we know ectogenesis is now universal" in such a way that only "less than 30 percent children are now born of woman" (Haldane 1923, 16). Haldane further advocates for the separation of sexual love from reproduction and applauds the method of selective breeding that was made possible "thanks to the work of Thomas Morgan on heredity substance". Haldane's ultimate aim is to change the character of human being "as quickly as institution". Haldane further writes,

> Because of the method of selective breeding . . . the small proportion of men and women who are selected as ancestors for the next generation are so undoubtedly superior to the average that the advance in each generation in any single respect, from the increased output of first class music to the decreased convictions of theft, is very startling. Has it not been for ectogenesis there can be little doubt that civilization would have collapsed within a measurable time owning to the greater fertility of the less desirable members of the population in almost all countries.
>
> (Haldane 1923, 166)

Haldane further speculates if the election placard in 300 years would deploy, "vote for Smith and more musicians", "vote for Macpherson and a prehensile tail for your great-grand children". Haldane thus anticipated *The Brave New World* of human beings so "bred" in the fluid outside of female bodies (Haldane 1923, 17).

More importantly, this projection of the "future possibilities" was founded on the reductionist theory of the particulate gene. At the time when very little of the material and physical or structural gene was known, the science was not only projected a couple of centuries ahead in the future but human immortality was being imagined in an entirely new way. Eugenics in the interwar period was not only envisaged as a sociopolitical paradigm, a marker of social and cultural inequality, but also as a tool for achieving human immortality. These views might sound truly bizarre and also politically incorrect from the location of our present in the first quarter of the twenty-first century, close to 100 years, if not 150, in Haldane's future; however, at that time such immortality ideologies had a wide currency, many scientists with progressive views believed in its potential to take humanity to newer heights of temporal and spatial expansion.

Such future speculations were founded on the metaphysical assumptions of the entities such as the heredity substance as orderly, fully law-governed, and hence predictable and controllable. In the first quarter of the twentieth century, the validity of the claims of the existence of immortal spirits and souls were questioned on the one hand, and on the other, the basic matter like atoms and molecules were imbued with eternity. For example, in 1909 Henry Frank in his book on science and immortality claimed how the vital matter "transmitted to our personal organism from the first bit of living substance that was formed on this planet" has never known death (Frank 1909). Frank further posits that "immortality is possible" when the organized intelligence would be able to control and dominate this (indestructible and primordial) vital matter. The issue of the indestructibility of the atom also played out in the debate between those scientists who believed in the psychical presence of the soul and spirit and those who did not. "Show me a single atom destroyed or built up – point to me a shred of proof worthy the consideration of a scientist that any element has ever been something other than it is today" (Kassel 1922, 524) asked a scientist challenging the psychical research. A number of such publications came out during this time that replaced the indestructibility and hence eternity of atoms with the eternity of the soul (Macfie 1909). The "matter" was made eternal and immortal by the intellectual climate that propounded the search of "single universal element, the primal stuff of all substance". The words of Solomon, acclaimed as the wisest man as per Judaism – "there is no new thing upon the earth", "the thing that hath been is that which shall be", and "all novelty is but oblivion" – were often quoted in these arguments (Pratt 1891, 334). In these discourses, the importance of atoms and molecules as the basic units of the universe, the core of reductionism, was claimed in the language and semantics of eternity, indestructibility, and immortality.

The basic matter thus was imbued with the property of eternity and understanding and controlling this matter was envisaged as a step in the direction of human immortality.

Immortality ideologies

Theory of the particulate gene and eugenics: Hermann J. Muller

Hermann J. Muller was born in 1890 and grew up in a lower-middle-class German and Irish neighbourhood in Manhattan. As per Muller's own autobiographical notes that he left for the National Academy of Sciences and also narrated to his friend and colleague, the Russian scientist Vavilov, he was brought up in an environment that imbued "a strong sympathy for the working class, for oppressed peoples, and for internationalism, together with a skepticism for the righteousness of established governments and a hatred of imperialism" (Carlson 1981, 15). Muller's biographer Elof Axel Carlson describes him as a lifelong idealist in the manner that he believed in an idealized future for humanity and that for the whole of his life he staunchly adhered to his belief in a certain socialist political ideal. Some of these beliefs, for instance, the belief in science's role in expanding humanity's spatial and temporal boundaries were adopted even before he started his career as a scientist. This idealism, however, often turned into controversial positions, for example, his lifelong belief in what he called the positive eugenics, his faith in communist Russia, and his claim of Jewish ancestry. During the high time of Nazism in Germany, Muller decided to claim his remote Jewish ancestry, which, as per his sister, was only a reaction to anti-Semitism, an act "to be", as his sister described it, "fully on the side of his persecuted friends". His sister thought that Muller's difficulty in finding a job in the 1940s was due to his self-proclaimed Jewishness, while, according to her, "there is NO relative on either side of our family who is considered Jewish by society" (capitalization original) (Carlson 1981, 14). Until he was seriously disillusioned with Stalinist Russia, Muller was also an ardent Marxist. He moved to Russia at the helm of his career with a hope that Stalin would implement his system of positive eugenics. In more than one way, Muller was willing to suffer a personal and professional death for his political idealism, and his conviction about positive eugenics, the most controversial of all, was one of them.

Muller advocated selective breeding for humanity while he was still an undergraduate student. In 1910, when he was 19 years old, Muller was invited to give a general talk on the subject of biology in the Peithologian Society, a literary and philosophical discussion group mainly attended by the students of humanities. He presented his thoughts on "Revelations of Biology and Their Significance", which was largely an extension of the textbook ideas he read in his biology course; however, in his lecture, he also included thoughts on eugenics such as, "the way to eliminate the unfit is to keep them from being born" (Carlson 1981, 34). Carlson thinks that at the time of delivering the lecture Muller must have been aware of (and possibly also influenced by) the eugenics movement founded by Francis Galton, but his own thoughts significantly differed from the caste-like elitism of Galton's eugenics movement. From the time of this undergraduate lecture, he related eugenics with "perfecting human evolution" – "work positively for the production of a nobler and nobler race of beings" (Carlson 1981, 34). And this, he clearly thought, could be achieved with science. It could be interpreted that Muller was merely expressing the dictum of the Enlightenment when he declared early in his career that "with knowledge of the laws of nature, comes power to manipulate them, and knowledge of life thus means the perfection of man" (Carlson 1981, 35). It is the argument of this chapter that Muller's idealism to "prefect" humanity's future became a powerful guiding principle for his choice of a particular path for his scientific research. Subsequently, in his biography, Carlson painstakingly shows that Muller's positive eugenics rejected the racist and classist underpinnings of the American eugenics movements and also the one founded by Galton, however, what is important for the arguments in this chapter is the fact that early on, even before Muller started his career as a scientist, his ideas on scientific exploration of heredity, gene, and mutation were applied to the imagination of human betterment. He in fact defined his positive eugenics system as "simply that which increases happiness and the human attributes that he believed would generate that happiness were sympathy, love, reason, and cooperation". Carlson describes this idealism as an "inner necessity to relate science to society" (Carlson 1981, 36).

In the context of the collective and individual affective climate in which the scientific truth-seeking was a struggle for immortality that Hermann J. Muller's founding work on the particulate gene unfolded. In Thomas Hunt Morgan's lab at Columbia University, where 22-year-old Muller began his career in 1912 as a biologist, the chromosome theory of inheritance was established in the combined work of Morgan, his students, and collaborators Alfred H. Sturtevant, Calvin B. Bridges, and Muller (Kohler 1994). The group worked on the common fruit fly, *Drosophila*. The chromosome theory propounded that the heredity factors were located on chromosomes close to one another in a linear series, like beads on a string, and their relative location and association in chromosomes were transmitted from one generation to the other. And, the physical exchange – crossover – between parts of chromosome accounted for mutations. At this time not much was known about the material – physical and chemical – basis of the gene (Kohler 1994). In fact, Morgan in his 1934 Nobel Prize lecture indicated that it did not make much difference if the gene was a real material particle or a hypothetical unit of measure. Contrary to the views of his teacher, Muller was determined to make the gene a physical reality, a thing that in his prophetic words, one could "grind in mortar and cook in beaker" (Muller 1922, 49).

Muller's work on systematic examination of relationship between a gene and phenotype character or trait was the beginning of a new chapter in the history of genetic science. At that time, although, Francis Galton's eugenics theory of genotype-phenotype, i.e., one-to-one relation between a gene and its trait, had started to find a wider societal and scientific popularity. However, contradictory opinions prevailed. Wilhelm Johannsen, who in fact introduced the term gene in 1909 conducted a study on beans and concluded that the variability in size of beans in a pure line – descendants of a single self-fertilizing plant – was due to environmental effects while they all inherited fixed and unchanging genotype (Schwartz 2008, 201). Johannsen thus refuted the genotype theory of phenotype. Muller with his close friend and collaborator Altenburg conducted a study in 1912–13 on *Drosophila* with truncated wings to show that a presence of at least one gene was needed to cause truncated wings. With the conclusion of this laborious and rigorous experiment the theory of stability of the gene and its necessity for the phenotype trait was initiated, which fundamentally challenged the environmental theories of phenotype expression (Altenburg and Muller 1920).

However, it was not only the empirical triumph of the study that Muller and Altenburg were concerned about. It was the "futuristic possibility" that this study opened up was most exciting. In the truncated wing work published in 1920 in the journal *Genetics*, Muller and Altenburg boldly anticipated application of the theory of stability of genes to controlling human psychological characters in future. They wrote that the psychological characters were "plastic, obscure, and complicated" in such a way that it would seem next to impossible to analyse them by the Mendelian factors alone and the only way to study them would be to develop a method of what they called "linked identifying factors", i.e., genetic basis on the chromosome (Altenburg and Muller 1920, 51). The complete realization of their laborious empirical work on the truncated wings in *Drosophila* was invested in speculating such future possibilities. As Heidegger argued, the logical science – empirical results and outcome – is necessarily oriented towards existential science, i.e., the scientific possibilities directed ahead in the future. And integral to such futuristic possibilities is the fact that "care" is the basic state of Dasein, which translates in Muller and Altenburg's wishes to improve the lot of humanity. Dasein's way of being-toward-death is being ahead of itself, future-oriented.

The crucial question, however, is not that the empirical results were projected into developing future potential of humanity, but that such speculations of future created conditions for specific empirical questions to be researched, illuminated, and conceptualized. The speculations for the futuristic possibilities were motivational in founding the reductionist concept of the gene in Muller's work. After establishing validity of the chromosome theory of stability of genes, Muller focused on understanding change in genetic structure, i.e., how it happens and how could it be artificially induced. The theory of such artificially induced change would make it possible to realize the "speculated" future. It was the theory of mutation that accounted for the theory of change in the genetic structure on which Muller next cast his eyes. Here, the temporal scale of future reference was no less than the entire human evolution. "Mutations that occur in nature are rare, 1 in 50,000 flies shows a mutation, and still the rate and incident of evolution depend on mutation", wrote Muller to Julian Huxley in 1919 (Schwartz 2008, 221). According to Muller a typical gene underwent mutation every 2,000 years and the fact that the gene was so stable meant that external factors cannot be accounted for in the rate of production of mutation. Muller further posited that even when the structure of a gene changed due to random mutation and correspondingly its function also changed, the "autocatalytic" property, i.e., the property of the gene to copy and reproduce itself, did not change. Muller initially expressed astonishment acknowledging this property of the gene, which eventually became one of the central aspects of his theory (Deichmann 2004, 213).

In his paper "Variation Due to Change in the Individual Gene", Muller tied this autocatalytic property of the gene to inherit variation (Muller 1922). He argued that inheritance by itself leads to no change, and variation leads to no permanent change. It is not inheritance *and* variation that bring about evolution but inheritance *of* variation (Carlson 1981, 128). The autocatalytic property of the gene, to repeat, the ability of the gene to "inherit the variation" – or, in other words, the property of the gene to copy and reproduce itself despite the "change after change" in its character – was then understood by Muller as a universal feature of the gene which he attributed to its physical and chemical structure (Muller 1922). The reductionist concept of the gene from the beginning is closely intertwined with the concepts of permanence and (almost eternal) reproducibility, which are since defined as the most central characters of the gene. In August

1926, at a meeting of the International Congress of Plant Scientists, Muller made the gene even primordial, he extended the concept of the gene to the origin of life itself and propounded a dictum "every gene from a pre-existing gene" and extended it to cover the entirety of life – "life itself originates as a naked gene" (Carlson 1971, 163). This theory of the gene opened up a whole new way of imagining the physical fate of humanity. Muller's gene-work began with projecting the fate of humanity in the future. This futuristic idea was further made into a political manifesto in his doctrine of positive eugenics which I discuss later.

Controlling and inducing mutations through the application of X-ray was Muller's next project. The X-ray work first fuelled Muller's soul-belief in eugenics, i.e., controlling human evolution to the desired end. But his wish to master the evolution was soon squashed only to be resurrected later in a different avatar. The X-ray research began in 1926 and within a few months Muller found more mutations than were found in all previous 16 years of work on natural mutations in *Drosophila* (Crow 1997). Once the possibility of X-ray-produced mutations was made public, Muller became an instant celebrity in America – the human beings had for the first time wilfully modified the genetic material (Schwartz 2008, 239). It was heroic because controlling mutations and hence controlling human evolution was now possible.

Holding such hopes, Muller exuberantly discussed the outcome of X-ray mutations and its philosophical and existential implications for the human evolution in a public lecture which was subsequently published in 1929 (Muller 1929). This is a brilliant essay in which Muller expertly covers a vast ground. He effortlessly merges the images of eternity of universe with the microscopic world of the gene, and ultimately projects the empirical results on the X-ray work onto controlling human evolution. He begins existentially, "for ages we have remained cooped up within blank walls of artificial construction, knowing of nothing beyond, behind, beneath, above", he then optimistically announces, "through a combination of luck and a certain amount of scheming, we have managed to creep through a series of crevasses in the structure, and gained the view for the first time of the brilliant real world outside". This "brilliant view" of the outside world has expanded "our universe" and hence "unlimited vistas have opened up", informs Muller. What grants the "brilliant view" of the outside for Muller is "discovering microscopic realm within realm in the interior of all parts of all things" (Muller 1929, 481). Muller here anchors the microscopic view with opening up the possibility to know the universe and thereby generating a promise of transcending the existential physicality of humanity. In other words, for Muller, the universe was made available to us to predict and control through the exploration of the microscopic world of the molecule of gene. Further in the essay, Muller discusses possibilities for the empirical X-ray work: What is the cause of mutations? The mutations caused by natural selection were rather scattered, sporadic, and rare and were not a result of molar processes but were caused by individual ultramicroscopic accidents. He also emphasizes the basis of his own gene theory, that is, when mutation occurred in one gene, the hundreds or thousands of other genes in the given cell remained unchanged. Then he describes the basic method of mutations induced in *Drosophila* through the application of X-ray radiation, and describes the results as "startling" and "unequivocal". He sounds ecstatic describing the feeling in these words, "to the toiling pilgrim after plodding through the long and weary deserts of changelessness, here indeed was the Promised Land of Mutations. . . . The roots of life – the genes – had indeed been struck, and had yielded" (Muller 1929, 491).

However, this sense of triumph soon turns into cautious warnings as the promise of the Promised Land, the desirable mutation, was still far away. Soon Muller reveals in the essay that although life was built of mutations, a greater number of natural mutations found in *Drosophila* were lethal, i.e., their effect was to kill the organism before it acquired maturity. Muller tells us that similarly the artificially induced X-ray mutations were in some way not only detrimental to the organism, but they also rarely produced desired traits, "they occurred without reference to their consequences" (Muller 1929, 493). That means that not only that "the change" induced by X-ray, "occurred accidentally, without reference to advantage or disadvantage it would confer", but most importantly, "they were steps in the wrong direction in the struggle for existence" (Muller 1929, 493). He also refers to his and Altenburg's earlier work to state that X-rays not only brought about changes in individual genes but caused breakages of entire chromosomes or string of genes, accompanied by reattachment of the broken-off fragments to different chromosomes. Muller calls this phenomenon of accidents and chaos a "molar indeterminism" (Muller 1929, 498). But, despite this acknowledgement of indeterminism, his prophetic, ecstatic mood is hardly dampened.

Further in the essay, he takes a refuge in an elaborate mathematical theory, finds a numerical pattern to account for the role of multiplication and selection (of mutations), and turns the lethal chaos of accidental mutations into a hypothetical order of advantageous changes. Muller postulates that the only way good mutations could be made to sustain is to make them multiply by means of selecting various individual good mutations and either combining them through sexual reproduction into one line of descent or by eliminating the unfit ones. And that is how out of the empirical work of

X-ray mutations Muller substantiates his political system of positive eugenics. The essay ends with a prescriptive statement, "Man must eventually take his own fate into his own hands, biologically as well as otherwise, and not be content to remain, in his most essential respect, the catspaw of natural forces, to be fashioned, played with and cast aside". Muller sounds even fanatic, as he proposes to take our fate into our hands and in doing so "we . . . must die if need be, with our eyes open" (Muller 1929, 505). This essay on the X-ray mutations and the method of evolution demonstrates how Muller's soul-belief – generating collective immortality – provided a powerful searchlight for his research while he groped in the darkness of indeterminate empiricism.

In Muller's representation of his X-ray work and its existential (im)possibilities I see an interplay of the rational and the irrational. The irrational here is not, as Otto Rank clarified, the lack of reason or madness, but the profound realization, however unconscious, that the events in the world are finite, inexplicable, and uncontrollable. In fact, in the beginning of the 1929 essay "The Method of Evolution", there is an element of "omnipotence" in Muller's grandiose prediction of control over the natural world. This turns into a sober mood when it is realized that X-ray cannot produce desirable mutations. Muller creatively deals with this irrationality – the profound awareness that nature is not that easy to control or master simply by an application of rational science. He turns irrationality of unpredictability into rationality by elaborate mathematical speculations and by proposing the societal program of eugenics based on such mathematical predictions. Muller's proposal of good mutations to be found and multiplied eventually developed into a full-fledged societal program of positive eugenics that he outlined in his 1936 book *Out of the Night: A Biologist's View of the Future* (Muller 1936).

Muller, like Borges' Flaminius Rufus (the mortal turned immortal) struggles to find purpose and order in the method of evolution. The impossibility to control the gene mutations spoils the "[orderly] view of the universe". In Borges' words such lack of order and pattern "pollutes the past and future and somehow compromises the stars". Exploring the method of evolution, Muller often arrives in the Borges' *City of Immortals* with "upside-down staircases . . . arriving nowhere", "corridors leading nowhere", "dramatic doors opening into empty shafts" (Borges 1998). For Muller, as for Rufus, as long as the city of purposelessness endures, no one in the world can be happy or courageous. Muller's system of positive eugenics – the one that "simply increases human happiness" – is founded on the purposefulness of order and pattern discovered in the microscopic world of the gene mutations. And Muller defying mortality finds this order and pattern one way or another, if not in the X-ray produced mutations, then in mathematical reasoning and in the societal program of positive eugenics.

Muller's science was driven by his unwavering belief in the eugenics and political idealism. So strong was this belief that Muller was even willing to die for it. The strength of his idealism was expressed in a letter to his to-be-wife at a delicate moment in their relationship, "it would not take much more to knock all idealism out of me, and for me that means death – virtually or actually" (Schwartz 2008, 234). His belief in eugenics shaped his complex relationship with his first wife driving him to attempt suicide; the same belief again put his life in danger when he stayed longer in Russia hoping that Stalin would implement his eugenics project while three of his close friends and colleagues were put in prison and one of them was executed for believing in the chromosome theory of the gene (Schwartz 2008, 267). Muller was willing to die for his immortality ideology. Ernest Becker quoting Otto Rank defends truth-seeking as immortality ideology by pointing out that all ideological belief and disputes amount to life-and-death viciousness. "Each person nourishes his immortality in the ideology of self-perpetuation to which he gives his allegiance; this gives his life the only abiding significance it can have", "he is even willing to die perpetuating his ideology", says Becker (1975, 64).

In the post–World War II period, eugenics is looked upon by most intellectuals as politically and ethically undesirable, but in the interwar period a significant number of intellectuals and life scientists earnestly believed in it. Muller, however, scathingly criticized the negative eugenics which he called sexist, racist, and based on the myth of the Nordic superiority. He openly wrote against the American eugenics movement, Nazism, and inequalities in capitalism (Carlson 1981, 6). In his book *Out of the Night*, Muller lays out a detailed plan on how to promote selected genetic traits that were assured to benefit humanity. This could be summarized in one line – creating societal means to control, what he called, indiscriminate reproduction. In other words, to discourage some individuals to have fewer children and others more (Muller 1936). Some of Muller's suggestions sound outrageous today, and in fact later in his life, he was indeed ridiculed and denigrated for holding them; despite this humiliation, he never abandoned his belief in eugenics (Carlson 1987). He suggested universal use of contraception in marriage relationships and artificial insemination of women from the seed of outstanding men only. He further suggested that for the artificial insemination the sperm of the greatest men only of the previous generation be used, i.e., only once the greatness of these outstanding men is attested by history – as the men believed to be great in this generation may not be held in the high esteem in the next generation. For the ethnic and racial diversity, he suggested a list, for example, of great men such as Lenin, Newton, Leonardo, Pasteur, Beethoven, Omar Khayyam, Pushkin, and more. "How many women", Muller exclaimed, "in an enlightened community devoid of superstitious taboos and sex slavery, would be proud and eager to bear a child of Lenin or Darwin!" (Muller 1936, 152–153).

Muller even zealously implemented these views in his own life and almost died bearing the emotional brunt of such a eugenics-induced, emotionally tumultuous marriage. His relationship with his first wife Jessie Jacob, a mathematician, was formed with such an ardent belief in eugenics. Although both their political beliefs matched well, from the beginning their emotional connection was fragile. Despite emotional volatility, the couple got married in 1923 because they believed that a child born to two highly intelligent, socially progressive, and atheist individuals was the best hope for the improvement of the gene pool. Jacob gave birth to their son in 1924 who was named as David Eugene – the middle name reflected their mutual belief in eugenics. Both Jacob and Muller also adhered to the principles of free love and open marriage so common among intellectuals at that time (Schwartz 2008, 236). Six years later, in 1931, Jacob had an affair with Muller's colleague and assistant Carlos Offermann, a new arrival in the lab, which devastated Muller and drove him to attempt suicide. He was found roaming in wilderness bruised and incoherent two days later (Schwartz 2008, 248). The suicide note, however, contained the detailed instructions for his friend Altenburg regarding the publication of his book on eugenics *Out of the Night: A Biologist's View of the Future*. So staunch was his belief in eugenics that later when the couple filed for the divorce and was fighting for the custody of their son, Muller wrote to Jacob, "if people who have so much in common as you and I have to use force and compulsion on one another instead of reasoned agreement and mutual concession . . . I think human nature is sadly failing". More than ever he now believed that the only hope to improve the human nature was eugenics, "we may some day make better people than the best we know" was his conclusion derived from this very intimate and deep personal crisis (Schwartz 2008, 257).

Conclusion

Muller's pioneering work on the particulate gene, thus, cannot be understood without his lifelong belief in controlling human evolution through eugenics. Firstly, it can be argued that there was isomorphism between the empirical outcome of his science and the immortality of the human self thus existentially/affectively desired. Muller early in life acquired his ideological moorings which not only remained unchanging through-out his life but also acted as a powerful guiding principle, some kind of psychic gyroscopes intimately shaping his science. Muller believed that knowing the realm of the microscopic world that the universe was given to humanity. Also, I show how Muller's science was directed ahead of itself. It was projected towards the field of possibilities of expanding the boundaries of human capabilities in the future by controlling evolution. For his theory of eugenics, the reductionist theory of the particulate gene was absolutely a prerequisite. And the obverse was true too – all his empirical results on the particulate gene, irrespective of their con-sequences, eventually culminated into his belief in eugenics. In fact, his soul-belief – generating collective immortality – provided a powerful searchlight for his research while he otherwise would have groped in the darkness of indeterminate empiricism. This way, I do not wish to argue for the causality between the affective and cognitive worlds, but the equiva-lence and consistency between their aims and forms. Lastly, it can be interpreted that in expressing his science and affective world view that Muller asserted his existential authenticity and his longing to find an immortal place for himself on the Earth.

Notes

1 The concept of isomorphism between the affective and intellectual paradigm comes from Lewis Feuer discussed in Söderqvist (1996, 69). Feuer in fact asks what is the *emotional a priori* of the scientist, what kind of the world that the scientist on emotional grounds sought to realize in the scientific theorizing? The term isomorphism is further debated in the conclusion.
2 For further discussion on the comparison of Hobbes' and Thucydides' immortal-ity ideology, see Ahrensdorf (2000).
3 For the further discussion on immortality ideology of Otto Rank, see Sheets-Johnstone (2003).
4 I need to clarify without making this chapter a heavy debate on Heidegger that Dasein means existence in German. In Heidegger, it is meant to refer to the pecu-liarity of human way of existing. Dasein is not a subject in the sense of having mental states and experiences, for Heidegger being is found not in thinking nature but in our existing in a world, a non-cognitive mode of being-in-the-world. For further discussion, see Dreyfus and Wrathall (2005, 3–5).
5 Heidegger (1985, 316–317) as quoted in Hoffman (1993, 195).
6 Heidegger (1962) as discussed in Hoffman (1993, 195–198).
7 Heidegger (1962, 293) as discussed in Rouse (2005, 176–177).
8 Based on the discussion in Rouse (2005, 175–187).
9 As discussed and quoted in Becker (1975, 64).
10 Rank (1998, 60). For a comparison of immortality ideologies of Descartes, Der-rida, Heidegger, and Otto Rank, see Sheets-Johnstone (2003).
11 For the biography of Haldane and his science and socialism, see Werskey (1978).

References

Ahrensdorf, Peter. 2000. "The Fear of Death and the Longing for Immortality: Hobbes and Thucydides on Human Nature and the Problem of Anarchy." *The American Political Science Review* 94 (3):579–593.

Altenburg, Edgar, and H. J. Muller. 1920. "The Genetic Basis of Truncate Wings: An Inconstant and Modifiable Character in Drosophila." *Genetics* 5:1–59.

Bauman, Zygmunt. 1992. *Mortality, Immortality and Other Life Strategies*. Cambridge: Polity Press.

Becker, Ernest. 1975. *Escape from Evil*. New York: The Free Press.

Borges, Jorge Luis. 1998. "The Immortals." In *Collected Fictions Jorge Luis Borges*. New York: Penguin Books.

Carlson, Elof Axel. 1971. "An Unacknowledged Founding of the Molecular Biology: H. J. Muller's Contribution to Gene Theory, 1910–1936." *Journal of the History of Biology* 4 (1):149–170.

Carlson, Elof Axel. 1981. *Genes, Radiation, and Society: The Life and Work of H. J. Muller*. Ithaca: Cornell University Press.

Carlson, Elof Axel. 1987. "Eugenics and Basic Genetics in H. J. Muller's Approach to Human Genetics." *History and Philosophy of the Life Sciences* 9 (1):57–78.

Crow, James. 1997. "Seventy Years Ago: Mutation Becomes Experimental." *Genetics* 147:1491–1496.

Deichmann, Ute. 2004. "Early Responses to Avery et al.'s Paper on DNA as Hereditary Material." *Historical Studies in the Physical and Biological Sciences* 34 (2):207–232.

Dreyfus, Hurbert, and Mark Wrathall. 2005. "Martin Heidegger: An Introduction to His Thoughts, Work, and Life." In *A Companion to Heidegger*, edited by Hurbert Dreyfus and Mark Wrathall, 1–17. Oxford: Blackwell Publishing.

Frank, Henry. 1909. *Modern Light on Immortality*. Boston: Sherman, French and Co.

Haldane, J.B.S. 1923. *Daedalus, or, Science and the Future*. A paper read to the Heretics, Cambridge, on February 4th, 1923 http://www.unife.it/letterefilosofia/lm.lingue/insegnamenti/letteratura-inglese-ii/programmi-anni-precedenti/programma-desame-2011-2012/J.%20B.%20S.%20Haldane-%20Daedalus-%20or-%20Science%20and%20the%20Future-%201923.pdf/view. Accessed on 12 April 2018.

Heidegger, Martin. 1962. *Being and Time*. Translated by J. Macquarrie and E. Robinson. New York: Harper Row.

Heidegger, Martin. 1985. *History of the Concept of Time*. Translated by Theodore Kisiel. Bloomington: Indiana University Press.

Hoffman, Piotr. 1993. "Death, Time, History: Division II of Being and Time." In *The Cambridge Companion to Heidegger*, edited by Charles Guignon, 195–214. Cambridge: Cambridge University Press.

Kassel, Charles. 1922. "Immortality and the New Physics." *The North American Review* 216 (803):523–534.

Kohler, Robert. 1994. *Lords of the Fly: Drosophila Genetics and the Experimental Life*. Chicago: University of Chicago Press.

Macfie, Ronald. 1909. *Science, Matter and Immortality*. London: Williams and Norgate.

Muller, Hermann J. 1922. "Variation Due to Change in the Individual Gene." *American Naturalist* 56:32–50.

Muller, Hermann J. 1929. "The Method of Evolution." *The Scientific Monthly* 29 (6):481–505.

Muller, Hermann J. 1936. *Out of the Night: A Biologist's View of the Future*. New York: Vanguard Press.

Pratt, W. H. 1891. "Immortality in the Light of Modern Dynamics." *Science* 17 (436):333–334.

Rank, Otto. 1958. *Beyond Psychology*. New York: Dover Publications, Inc.

Rank, Otto. 1998. *Psychology and the Soul: A Study of the Origin, Conceptual Evolution, and Nature of the Soul*. Translated by Gregory Ritcher and James Liberman. Baltimore: John Hopkins University Press.

Rouse, Joseph. 2005. "Heidegger's Philosophy of Science." In *A Companion to Heidegger*, edited by Hurbert Dreyfus and Mark Wrathall. Oxford: Blackwell Publishing.

Schwartz, James. 2008. *In Pursuit of the Gene: From Darwin to DNA*. Cambridge: Harvard University Press.

Sheets-Johnstone, Maxine. 2003. "Death and Immortality Ideologies in Western Philosophy." *Continental Philosophy Review* 36:235–262.

Söderqvist, Thomas. 1996. "Existential Projects and Existential Choice in Science: Science Biography as an Edifying Genre." In *Telling Lives in Science: Essays in Scientific Biography*, edited by Michael Shortland and Richard Yeo, 45–84. New York: Cambridge University Press.

Werskey, G. 1978. *The Visible College: The Collective Biographies of British Socialists of the 1930s*. New York: Holt, Rinehart, and Winston.

4

WHAT IS LIFE? PLACED IN ERWIN SCHRÖDINGER'S LIFE

> We may safely assert that there is no alternative to the molecular explanation of hereditary substance. The physical aspect leaves no other possibility to account for itself and of its permanence.
>
> (Schrödinger 1967[1944], 57)

> Let us see whether we cannot draw the correct, non-contradictory conclusion from the following two premises: (i) My body functions as a pure mechanism according to the Laws of Nature. (ii) Yet I know, by incontrovertible direct experience, that I am directing its motions, of which I foresee the effects, that may be fateful and all-important, in which case I feel and take full responsibility for them. The only possible inference from these two facts is, I think, that I – I in the widest meaning of the word, that is to say, every conscious mind that has ever said or felt "I" – am the person, if any, who controls the "motion of the atoms" according to the Laws of Nature. . . . In Christian terminology to say: 'Hence I am God Almighty' sounds both blasphemous and lunatic. But please disregard these connotations for the moment and consider whether the above inference is not the closest the biologist can get proving also their God and immortality in one stroke.
>
> (Schrödinger 1967[1944], 86–87)

The book *What Is Life?* made Schrödinger one of the founding fathers of molecular biology. It is often even projected as one of the most influential scientific books of the twentieth century (Fleming 1968). Gunther Stent, in his wryly written essay "Waiting for the Paradox", argues that although the ideas presented in the book were considered neither particularly original nor novel, it was so successful because it had a propaganda effect on physicists whose knowledge on biology was then limited (Stent 1966). As for biologists also equally impressed, the book combined physical science with life sciences, which was not so common those days. Several scholars have argued that the book provided much-needed metaphor and the rhetorical concept of the gene as code-script that was instrumental in inaugurating structural molecular biology in particular (Keller 2002; Judson 1979; Olby 1994; Doyle 1997). The origin of the concept of code-script is generally attributed to the war-born disciplines of cybernetics and information theory especially in the work of Norbert Wiener and Claude Shannon. However, the ideas presented in *What Is Life?* are also traced to the older discourses and representation of life sciences. The concept of *aperiodic crystal* discussed in the book is even traced to the seventeenth century (Kay 2000, 61). Nonetheless, *What Is Life?*'s generative role in shaping molecular biology and Schrödinger's founding role is only rarely disputed in the history of molecular biology.

What is not discussed in this historiography of *What Is Life?* are two seemingly incompatible statements in the book – as in the two epigraphs quoted at the beginning of this chapter taken from the book. The first epigraph sets the overall tone of the book. Schrödinger argued in this book to make the gene a material reality, a stable molecule, which was subjected to change, i.e., mutations, due only to changes made in physical and chemical structure. Further, this material/molecular gene was the substance that accounted for the continuity and permanence of life itself. *What Is Life?* made the gene a researchable entity by emphasizing materiality, stability, and permanence of the gene. This was reductionism par excellence and arguably the most important reason for the book's wider appeal to both physicists and biologists in the aftermath of World War II (Olby 1971; Domondon 2006). It is beyond dispute that the main argument in *What Is Life?* is about scientific materialism. Schrödinger here sounds like a mechanist, comparing the gene with the clock.

The second epigraph quoted at the beginning of the chapter is taken from the concluding chapter of *What Is Life?*. While the rest of the book is composed of lectures given at Trinity College in Dublin in 1943, the last chapter, called epilogue "On Determinism and Free Will", was added only at the time of publication. In the epilogue, Schrödinger expresses his otherwise lifelong belief in the Indian philosophy of Vedanta, lashes out at "Official Western Creeds" for their "gross superstition" for believing in individual and plurality of souls, and argues for unity or oneness of consciousness and eternity of the human Self (Schrödinger 1967[1944], 88). Not only that the gene, the molecule, is stable and permanent, but "I", who control the motion of atoms (in the molecule), is part of an all-encompassing singular consciousness and hence even "I" is immortal, beyond death. *What Is Life?* not only labours to establish permanence of the gene and hence life, but ends with the prophetic words of individual immortality. "In no case is there a loss of personal existence to deplore. Nor will there ever be" (Schrödinger 1967[1944], 90). In other words, the atoms in a molecule are permanent, and "I", who controls the atoms, too, is eternal. Schrödinger here aligns the individual, atoms, nature, and God qua consciousness in the string of immortality. But these ideas inspired by the Eastern mysticism are not compatible with the stark reductionism, physicalism, determinism, and materialism of the gene expressed in the rest of the book.

Compared to the rest of the book, the epilogue clearly looks like an appendage. So why did Schrödinger include the epilogue? Was it just an aberration, some kind of hobby that he pursued on the side? The epilogue was not part of the lectures but was added in the written version just before the publication, and therefore it indeed looks like an afterthought. But it was not included light-heartedly. It was included at the cost of losing a much-reputed publisher in Dublin, Cahill and Co. At the final stages of the publication the publisher raised objections to the ideas in the epilogue, especially to the scornful and derogatory remarks about the Church. Schrödinger's biographer Walter Moore thinks that Schrödinger was perhaps persuaded to delete the epilogue but he refused and at the last moment the publisher withdrew. The book was eventually published a year later by the Cambridge University Press (Moore 1989, 401–403).

Moore says that much of what Schrödinger said in the epilogue was repetition of what he wrote in *Meine Weltansicht (My World View)* 18 years ago in his personal journal (Moore 1989, 401). *Meine Weltansicht*, however, remains Schrödinger's first expression on the genetic inheritance and memory, which was written much before he wrote *What Is Life?*. What is important is that it remains the last word, too, because Schrödinger barely revised his views expressed in *Meine Weltansicht* in the version *My View of the World* published before he died in 1961 (the English translation was published in 1964) (Schrödinger 1964). The continuity of ideas between *Meine Weltansicht* and *My View of the World* makes it difficult to imagine that Schrödinger wrote *What Is Life?*. Some of these ideas were also elaborated in the text *What Is Life? with Mind and Matter* published in 1958 – a sequel to *What Is Life?*. This was also a collection of lectures he had delivered at Dublin in 1956 (Schrödinger 1967).

Not only that these metaphysical ideas left the younger generation of scientists so impressed with *What Is Life?* relatively uninterested (Teich 1975, 277), but also that these ideas are rarely discussed or mentioned in the historiography of the importance of *What Is Life?* in molecular biology. The historiography of *What Is Life?* has strictly engaged with the reductionism in the book and has ignored Schrödinger's philosophical and metaphysical contributions and thoughts that do not fit in these boundaries. The purpose of this chapter is to discuss the continuity of ideas expressed in the epilogue in all four texts – Meine Weltansicht, What Is life?, What Is life? with Mind and Matter, and My View of the World – and to show that the metaphysics expressed in these ideas remained an important backbone of Schrödinger's life and thought. The question I ponder upon is: Why was the famous physicist interested in the question of the gene and heredity? My claim is that placing What Is Life? diachronically in the context of Schrödinger's life and thought creates an entirely different picture about his motive for writing this book than the one that is usually painted in the history of molecular biology and historiography of What Is Life?. My ultimate aim is to contribute to the debates on who is the scientist-subject doing science and what is science and how it is done.

Historiography of What Is Life?

Two issues are generally discussed in the historiography of What Is Life? – firstly, the popularity of the book among generations of molecular biologists, and secondly, its contribution to defining the future course of molecular biology. Robert Olby, in his extensive review, places the book in the scientific culture of the 1940s and 1950s when the majority of the scientists were reductionists. Olby, however, thinks that the emergence of the quantum physics in the 1950s and 1970s supplied a major backlash against the "overconfidence and intellectual despotism of the reductionist approach" (Olby 1971, 120). Schrödinger himself being one such anti-reductionist quantum physicist, Olby wonders "what prompted Schrödinger to write What Is Life?" in this intellectual climate (Olby 1971, 121). Although Olby mentions Schrödinger's other intellectual concerns such as "mind and matter", "free will", and "nature of the soul" and counts "the nature of consciousness and free will" as one of the four questions Schrödinger aimed to address in What Is Life?, he ultimately restricts his own review to three other questions that he thinks are probed in What Is Life?: How organism resists destruction, how hereditary substance remains unchanged, and how this substance reproduces itself. Restricting his review to these issues, Olby chooses to project that Schrödinger in What Is Life? was chiefly concerned with the problem of reduction. Olby even goes a step further to argue that by not addressing the fourth question on the nature of consciousness and free will "the problem of reduction in biology is made far less daunting" (Olby 1971, 125).

In the similar vein, reviewing Olby's other book *The Path to the Double Helix* (Olby 1994), historian Makulas Teich first asserts that "Schrödinger's book *[What Is Life?]* cannot be separated from the sometimes heated controversies . . . on the perennial problems of chance and necessity, determinism and free will, nature and God" (Teich 1975, 277). Teich, however, eventually argues that many of these philosophical questions that had excited Schrödinger, which were closely argued in *What Is Life?* in "readable and lucid fashion (except for the epilogue) . . . left younger generation relatively cold" (Teich 1975, 277). Teich also thereby not just places but also reviews *What Is Life?* with respect to its reductionism.

Olby and Teich count reductionism as the reason for the book's popularity among the younger generations of scientists. But, it is not only the generations of scientists but even historians have remained detached from exploring the metaphysical agenda in Schrödinger's work on biology. Yoxen wonders if the connection between physics and molecular biology so explored in *What Is Life?* extends to the metaphysical and philosophical issues mentioned by Teich (i.e., chance and necessity, determinism and free will, nature and God) (Yoxen 1979, 19). After consulting Schrödinger's notebooks and correspondence not previously discussed, Yoxen comments on Schrödinger's motivation and inspiration for writing the book. He thinks that the concept of the gene as code-script got undue importance in discussing the book's influence, whereas the real value of the book for many scientists lies in its contribution to the problem of order in molecular biology (Yoxen 1979, 26, 30). However, referring to notebooks that Schrödinger wrote at the age of 18, Yoxen does acknowledge Schrödinger's deep interest in Indian mystical writers, also philosophers like Spinoza and Mach, and, what Yoxen names as "philosophically oriented" scientists, Fechner and Semon (Yoxen 1979, 31). Further referring to Schrödinger's other work, Yoxen does mention that Schrödinger was thinking about ideas in *What Is Life?* at least ten years before the book was written and was passionate about it, on Schrödinger's own admission, "like old Goethe's heart was in the Farbenlehre" (Yoxen 1979, 35). While Yoxen does acknowledge that Schrödinger was passionately musing about the ideas in *What Is Life?* for most of his adult life and that these ideas were connected to his interest in Indian mysticism and the ideas on inheritance and memory in the work of Richard Semon, these are not the influences that Yoxen gives serious consideration in his own analysis. In a footnote, he even rejects the possibility that Schrödinger's idea of the hereditary code-script could have been influenced by Semon's work on *engrams* and *mneme* (Yoxen 1979, footnote 64). Indian mysticism and Semon are two influences on Schrödinger's life and thought that historians of molecular biology are not comfortable taking seriously. This way, as Olby stated, "the problem of reduction in biology is made far less daunting" (Olby 1971, 125).

Similarly, historian Andrew Domondon comparing the views of three leading physicists involved in the birth of molecular biology – Bohr, Delbrück, and Schrödinger – argues that "despite Schrödinger's undeniable interest in Indian philosophy, it is a mistake to suppose that it played a role in Schrödinger's worldview that is analogous to the role played by complementarity in the worldviews of Bohr and Delbrück" (Domondon 2006, 447). Schrödinger indeed states in *What Is Life?* that "contrary to the opinions upheld in some quarters, quantum indeterminacy plays no biologically relevant role" in the cell (Schrödinger 1967[1944], 86). In fact, Domondon may be correct in disapproving the comparison between Bohr's complementarity principle and Schrödinger's engagement with Indian mysticism and Richard Semon's work on memory. However, Domondon goes a step further and argues that the main philosophical perspective that dominated *What Is Life?* was determinism and that his views on Indian philosophy might not be that important because such references were made only in the last few pages (Domondon 2006, 447). Quite like Olby, Domondon also makes the problem of *What Is Life?* "far less daunting" by marking the ideas on the last pages as unimportant.

What Is Life? is often described as one of the texts that inspired the whole new generation of scientists who built the new discipline of structural molecular biology (Ceccarelli 2001, 63). But, neither Schrödinger's ideas on *mneme* as generational memory nor his mysticism figure in the rhetorical reading of the book. In fact, in a similar attempt to analyse the rhetorical origin of the new regime of molecular biology, Doyle finds it problematic that by calling the gene as code-script Schrödinger framed the question of life in a reductionist framework and located the secret of life in code structure and not in actual organism (Doyle 1997, 34). But, Doyle does not engage with Schrödinger's metaphysics in the epilogue of *What Is Life?*. Other historians with the only mandate to discuss the role of physics in the development of molecular biology obviously do not debate the influence of Schrödinger's metaphysics on his biology (Keller 1990). And those who do mention that Schrödinger wrote books about philosophy and about the mind, they treat these interests more like a hobby, quite like writing poetry, which Schrödinger pursued on the side, which, it is also assumed had very little to do with his science (Symonds 1986, 226).

What Is Life? placed in Schrödinger's life

Schrödinger's interests in the questions of heredity and living organism's memory were first developed when he was 19 years old when he read

Richard Semon's work on *Mneme* with his friend Franzel (Schrödinger 1967, 167). Schrödinger then retained a lifelong fascination for these issues. Apart from his scientific work, Schrödinger also significantly published on philosophy, mind, and matter. As I already mentioned, I will now closely engage with Schrödinger's four texts – *Meine Weltansicht, What Is Life?, What Is Life?* and *Mind and Matter*, and *My View of the World* – because they engage with the questions of memory and inheritance centrally explored in *What Is Life?*, including in the epilogue.

The main message of the three texts – the personal account *Meine Welt-ansicht* (*My World View*), the published version *My View of the World*, and the epilogue in *What Is Life?* – is an argument of unified consciousness, namely the subject-object, and the self-world, unity. It is often argued that Schrödinger rejected the Copenhagen interpretation of the quantum mechanics because of his belief in the unity of consciousness not only among the living but also between scientific and philosophical thoughts (Fischer 1984). This search for unity, however, began early in his life, which culminated into the personal account *Meine Weltansicht*. The year 1924–25 when the personal account *Meine Weltansicht* was written was a crucial year in Schrödinger's life. He was painfully aware that he was getting older and had not made any major contribution to physics. At that time, the age of 30 was usually considered the end of the creative age. This is expressed in a poem by his physicist friend Paul Dirac (Moore 1989, 157):

Age is of course a fever chill
That every physicist must fear
He's better dead than living still
When once he's past his thirtieth year

Having passed the 38th year instead, Schrödinger was facing the threat of creative death. Moore thinks that the personal account must have been a result of meditation over many days (Moore 1989, 168–176). In addition, his four- to five-year-old marriage with Annemarie Bertel was going through a tumultuous period. They had not yet had a child, and the sexual incompatibility between the couple was such that it was unlikely that the marriage would have yielded the offspring that Schrödinger so longed for. The couple also followed the free thinking so common among intellectuals at that time and followed their own separate intimate and sexual liaisons while still staying married. Despite the fact that Schrödinger had had sev-eral casual affairs by then, he was unhappy with his wife's evident interests in other men. The intense personal account was written at this unhappy moment in Schrödinger's life. Moore thinks that there is little joy and no love at all in *Meine Weltansicht* (Moore 1989, 158–159). In fact, there is only an abstract contemplation with two forms of immortality – permanence of inherited memory and eternity of unified consciousness – not just in *Meine Weltansicht* but in all four texts. The issue of his own somatic and spiritual inheritance – the desire to have an offspring (a son actually) – remained a difficult personal issue throughout the period in which these four texts were written. Schrödinger eventually fathered two daughters with two different women, but each mother decided to raise her child away from him. Both of his daughters did not know he was their father until they were adults (Moore 1989, 255, 293, 423, 443). The deeply unfulfilled longing to have a child to pass on his legacy remained an emotionally challenging issue throughout his life.

My View of the World

My View of the World that Schrödinger revised and published just before he died reflects the importance of ideas on memory and inheritance both in his personal life and metaphysical thought. In Chapter 5, Vedantic Vision, Schrödinger provides the background of his argument unfolding in the next chapter, An Exoteric Introduction to Scientific Thought, on genetic memory (Schrödinger 1964). Schrödinger first addresses the reader and states that

> the conditions for your existence are almost as old as the rocks . . . for thousands of years men have suffered and felt joy and experienced awe for the beauty of mountains and valleys and glaciers . . . a hundred years ago a man sat in the same place as yours and felt the same awe for the dying light on glaciers. Was he someone else? Was he not you yourself? What is this Self of yours?
>
> (Schrödinger 1964, 20)

Schrödinger then argues that it is not possible that this individual "unity of knowledge, feeling and choice" sprung in the being from nothing. They in fact are essentially eternal, unchangeable and numerically *one* not just in human beings but in all living beings. He further argues that the Man is not just a part or a piece of an eternal, infinite being, he is the whole. Quoting Vedanta, Schrödinger then declares, "I am in the east, and I am in the west, I am below and above, I am this whole world" (Schrödinger 1964, 21). Schrödinger began taking interest in Vedanta and Buddhist philosophy, especially in eternity of the Self, during the time when *Meine Weltansicht* was being written. He writes in his journal in 1918, "Nirvana is a state of pure blissful knowledge. . . . It has nothing to do with the individual. . . . When a man dies, his Karma lives and creates for itself another carrier". A few days later, he writes, "No self is of itself alone. It has a long chain of intellectual ancestors. The 'I' is chained to ancestry . . . this is not mere allegory but an eternal memory" (Moore 1989, 113). The themes of ancestry, inherited memory, and eternity of the Self repeatedly emerged in all four books we are discussing here, including in the epilogue of *What Is Life?*.

The next chapter of *My View of the World* makes several references to Richard Semon's work (Schrödinger 1964, 23–29), and so does *What Is Life? and Mind and Matter*, which I discuss later. Schrödinger here is fascinated by the process of inheritance, the genetic line extending through the evolution. He exalts the idea of immortality through the continuity of genetic material, "the acts of propagation by which a series of genetically connected individuals proceed one from another are not really an interruption but only a constriction of both bodily and spiritual life" (Schrödinger 1964, 23). Schrödinger thus begins with stating that in the act of reproduction a genetically similar individual is produced that is continuous with its ancestors and its predecessors in body and mind. What is interesting here is that he compares the awareness of an infant in relation to its ancestors with his own consciousness before and after deep sleep. In order to counter the usually made argument that the infant cannot have any memory of the ancestral events, he gives several examples of instinctive behaviour as supra-individual memory. For example, he argues that human sexual behaviour is unlearned and instinctive.

> There is one complex in man . . . which has a strong emotional coloring, and bears the unmistakable mark of supra-individual memory: the first awakening of sexual feeling, the feeling of attraction and repulsion between the sexes, sexual curiosity, sexual shame and so on.
>
> (Schrödinger 1964, 24)

His second example is *ekphoria* (*ecphoria*) or a primitive inherited *engram* (from Richard Semon). Schrödinger argues that the way in which an individual might be carried away by effects of the fight or flight reaction is mainly because of the instinctive forces and emotions. "The whole organism visibly prepares in an admirable fashion for what thousands of our ancestors in similar circumstances actually did: for physical attack or defense against the aggressor" (Schrödinger 1964, 25). It is in the continuity of these ancestral events that Schrödinger seeks to find eternity of the Self. He further discusses such continuity of inherited memory with examples of a variety of organisms. Primitive animals like *Hydra fusca* and *Planaria* must have some form of consciousness, if cut into two, they regenerate into new complete animals with the same consciousness of the original creature (Schrödinger 1964, 26). He extends this argument to human beings: How is my "Self" composed of the individual "Selves" of my brain cells? For Schrödinger, my Self and the Selves in my brain cells are one and the same – that is the exact message he tries to convey in the epilogue of *What Is Life?*. "I" that controls the motion of atoms and the laws of nature are one in the same. Schrödinger ultimately relates these arguments to his belief that all human beings together constitute a higher unity, analogous to the constitution of an individual man from his somatic cells. He writes, "my conscious self depends on a particular . . . functioning on the part of my soma . . . these are in direct causal and genetic dependence on the structure and way of functioning of earlier somata" (Schrödinger 1964, 27). Thus, Schrödinger argues that his body has memory received from ancestral bodies – not just those bodies genetically connected but also those that did not. This point repeats even in *Mind and Matter* that he wrote in 1958. Here he argues not only that there is only one mind and one unified consciousness but that consciousness is associated with *learning* (the form of memory) by living substance. In *Mind and Matter* he expresses this in terms of "The *knowing how* of life is consciousness" (Schrödinger 1967[1958], 99).

In *My View of the World*, the continuity and permanence of somatic qua genetic memory translates into eternity and indivisibility of consciousness (Schrödinger 1964, 23–29). Schrödinger further argues that the individual's course of development depends upon special arrangement of genes and special pattern of environment. He then thinks that special arrangement of genes and special pattern of environment are one in the same because genes develop only under the influence of environment. Here, the influence of Lamarck is clearly visible – yet another reason for historians to feel uncomfortable. But what is important is what he says next, which in fact means a deep influence of thinkers like Carl Jung, although he does not acknowledge so, not in this book. He repeats that structure of individual spiritual self is essentially a direct consequence of ancestral events, not necessarily the individual's own physical ancestors. In other words, the genetic qua spiritual personality is bound up with environmental influences which are the direct outcome of other somatic qua spiritual personalities, some living, others dead. This way, he ultimately compares genetic inheritance with spiritual inheritance – they in fact are one in the same. The genes are thus modified not just by somatic inheritance from direct physical ancestors but by the environmental influences which are in fact spiritual influences on the bodies by other bodies in such a way that there is no difference between spiritual influences and genetic influences of the physical ancestors. Schrödinger concludes this chapter by declaring that "No Self stands alone. Behind it stretches an immense chain of physical and mental events". The individual self thus is a reacting member, the one who carries on, passes on the spiritual and physical memory. In fact, in the strict sense, the Self is not just a product of ancestors but the continuation of THE SELF (my emphasis) – the SAME THING (Schrödinger's emphasis) (Schrödinger 1964, 25–28). Schrödinger here means that "I" is the whole in the sense that each one of us can say "L'etat, c'est moi" (I am the state). An individual's hopes and fears are the same as those of thousands who have lived before him and one can hope that the future centuries may bring fulfilment to an individual's yearnings of centuries ago – this is the statement with which Schrödinger concludes his chapter on genetic inheritance in *My View of the World* (Schrödinger 1964, 29).

Amazing that *My View of the World*, which was published the year Schrödinger died, has no trace of materialism and reductionism of *What Is Life?*. Schrödinger in fact writes in the foreword that this little book was really a fulfilment of a very long-cherished wish (Schrödinger 1964). My claim is that the influence was the other way round. The gene is nothing but physics and chemistry of *What Is Life?* happened because of Schrödinger's much-cherished wish to write about somatic and spiritual memory, its inheritance and its permanence.

What Is Life?

In my interpretation, stability and permanence are the ultimate keywords in which the gene is described in *What Is Life?*. And, physicalism and order become vehicles by which Schrödinger establishes these tropes of stability and permanence. Schrödinger first discusses how for half a century it was known that the chromosome was heredity material and that it was able to replicate, just how it replicated was the matter of speculation. He emphatically rejects the idea of some special non-physical force or energy, for example, entelechy, operating in the organism. Further on, to establish the importance of replication and repetitiveness, he calls the gene *aperiodic crystal*, which means that it had a regular array of repeating and still varying units. Francis Crick, in conversation with Horace Judson, described Schrödinger's science "embarrassingly gauche"; referring to the concept of aperiodic crystal he quibbled, "I don't suppose the man had ever heard of a polymer!" (Judson 1979, 245). Historian Robert Olby also contends that Schrödinger oversimplified the idea of the gene and deliberately ignored the covalent bond which unites neighbouring nucleotides into a polymer chain (Olby 1971, 132). Although

Schrödinger's biology sounds amateur, biology of a "naïve physicist" in his own words (Schrödinger 1967[1944], 19), Schrödinger introduces his idea of the chromosome fibre as a code-script by describing the gene molecule as made of repeating and still varying units in the form of aperiodic crystal. He argues that the chromosome fibre was a linear code that contained the information that ultimately constituted genotype of a particular individual organism. Although Schrödinger first deliberates on the materiality of the gene molecule, he ultimately wants to establish the stability and permanence of the gene molecule through replication even when it was subjected to change. Referring to Delbrück et al.'s paper, popularly known by then as the "green pamphlet", Schrödinger further discusses how mutation of the gene molecule by X-ray is due to the chemical change the radiation produced in the molecule. But it is the continuity of genotype over hundreds of years that Schrödinger wishes to explain. He ponders that the continuity of genotype can only be explained if the aperiodic crystal which carries the genetic code is a stable molecule. And, implied in this stability is the permanence of the genetic material. Even when there is a degree of tautology in Schrödinger's argument – the permanence can be explained only in terms of stability and stability means permanence, the question that Schrödinger asks was: How was this stability qua permanence achieved? He resorts to physicalism only to state that "the enigmatic biological stability is traced back only to an equally enigmatic chemical stability" and then he speculates that such stability is likely to involve other laws of physics hitherto unknown (Schrödinger 1967[1944], 47). No such new laws were eventually found, as we all know. In my reading, Schrödinger here asserts physical reductionism as the only alternative to explain the gene but this he does to ultimately account for the permanence and stability of the gene. Reiterating his words, "we may safely assert that there is no alternative to the molecular explanation of hereditary substance. *The physical aspect leaves no other possibility to account for itself and of its permanence*" (italics mine) (Schrödinger 1967[1944], 57).

My claim is that reductionism in *What Is Life?* was yet another medium through which Schrödinger aspires to deliberate on the permanence of the material of inheritance. In *My View of the World* it was the similar message of memory (somatic and spiritual) at the core of eternity (permanence) of the (one) Self, and in *Mind and Matter* Schrödinger takes recourse to Lamarck through Semon to show how the stability and permanence of heredity material leads to the path of perfection of human evolution (this I discuss later). To repeat, my claim is that the physicalism and reductionism in *What Is Life?* were tools or vehicles to drive the point that replication or repetition (of genetic material) at the heart of inheritance contributes to stability and permanence of living organisms. I do not mean to suggest that Schrödinger was any less convinced or concerned about the issues of physicalism, reductionism, and order in *What Is Life?*, but I mean to suggest that the metaphysical search for permanence, unity, and eternity of consciousness runs through all of Schrödinger's thoughts on heredity and the reductionism in *What Is Life?* was yet another (and not the only) expression of this search. I would go to the extent of suggesting that Schrödinger wrote *What Is Life?* because of his larger metaphysical interests in the question of memory and unity.

In fact, the epilogue was not an aberration or hobby, relatively unimportant compared to the rest of the text as many historians of *What Is Life?* have treated it. Historian Ernst Peter Fischer comments on the whole of Schrödinger's intellectual trajectory including his explorations in biology and his thoughts on *One Single Being*, and argues that the epilogue was the culmination of the most pressing philosophical question that haunted Schrödinger his whole life (Fischer 1984, 834). And I can't agree more. As a physicist, Schrödinger never sided with the uncertainty, randomness, indeterminacy, and discontinuity implied in the mechanism of quantum jumps. He visualized an electron not as a point mass but as a standing wave waning and waxing in the atom implying continuity. An electron as a wave changed its position from one mode of vibration to another by making such event continuous in space and time. Schrödinger thereby countered quantum indeterminacy with determinism of his wave theory. And, as I have already discussed, some historians do think that determinism, Schrödinger's lifelong project, was the main message of *What Is Life?* (Domondon 2006). As I have already quoted before, Schrödinger clarifies in *What Is Life?* "contrary to the opinions upheld in some quarters, quantum indeterminacy plays no biologically relevant role" in the cell (Schrödinger 1967[1944], 86). In *What Is Life?* Schrödinger indeed embarks on showing how the living cell has well-determined behaviour, how it functions like a clock, like a thermal machine. What is missed in Domondon's historical argument is Schrödinger's further rumination on the issue of free will in the epilogue of *What Is Life?*. If the events in the living cell are strictly deterministic, how to account for the immediate experience of free will? The question that Schrödinger asks is: How can the conscious "I" control something whose character is determined by the laws of nature? While in the entire book Schrödinger stays on the path of showing how the action of the living cell is determined by the laws of nature, in the finale of the epilogue he relates determinism with conscious free will. That is, the free and conscious "I" and the deterministic laws of nature are not incompatible, they are one in the same. "I" am the person who controls the motion of atoms as per the laws of nature and "Hence I am God Almighty" and I am immortal – is the crux of

Schrödinger's message here (Schrödinger 1967[1944], 87). The epilogue was not an aberration or an appendage; it was a culmination of Schrödinger's entire life's thought and his belief in unity and eternity of consciousness.

What Is Life? and Mind and Matter

In *Mind and Matter* Schrödinger approaches the question of inheritance from the perspective of the whole of evolution. He begins by exploring the material basis for consciousness. He again starts with the role of repetitiveness in biology and refers to Semon's *Mneme* to argue that repetition of the event is integrated in consciousness in a way that "the individual soma is the 'well memorized' repetition of a string of events that have taken place in much the same fashion a thousand times before" (Schrödinger 1967[1958], 95). "A single experience that is never to repeat itself is biologically irrelevant" (Schrödinger 1967[1958], 96). Following Semon, Schrödinger argues that these repetitive events integrated into memory eventually become hereditarily fixed (in the consciousness). Schrödinger then proposes that this way consciousness is the phenomenon in the zone of evolution. He brings the ontogeny and phylogeny in an interesting way together that echoes his fascination for continuity of somatic and spiritual memory and the oneness of the Self. He says,

> any individual life as a whole is but a minute blow of the chisel at the ever unfinished statue. But the whole enormous evolution we have gone through in the past, it too has been brought about by myriads of such minute chisel blows. . . . For we ourselves are chisel and statute, conquerors and conquered at the same time. It is a true continued self-conquering.
>
> (Schrödinger 1967[1958], 100)

The individual chisel strokes thus contribute to the statue of evolution whereas the (statue of) evolution makes the individual chisel strokes possible. Here, replication and repetitiveness are the source of heredity and permanence which culminates into the making of whole of evolution. Schrödinger extends this discussion on the material basis for consciousness to the issue of "path to perfection" for the biological future (Schrödinger 1967[1958], 115). He is, however, well aware that his conclusion that the individual chisel strokes could be incorporated in the evolution of human species was based on the postulate of "inheritance of acquired character" and hence on much discredited Lamarckism. He does acknowledge so and warns against sham-Lamarckism, although he also confesses that these considerations have remained with him for more than 30 years (Schrödinger 1967[1958], 102). He keeps pondering about the path to perfection and asks if still physical evolution be expected in man? In the section titled as "Apparent Gloom of Darwinism" he says the idea that small inheritable changes called mutations from which the profitable ones are automatically selected are only small evolutionary steps and have only a slight advantage. He finds the whole Darwinian idea of evolution, in which "the individual has not the slightest influence on the hereditary treasures he receives and passes on", as "depressing, gloomy and discouraging" (Schrödinger 1967[1958], 105–106). And then he calls the Lamarckian idea of *inheritance of acquired characters* as "beautiful, elating, encouraging and invigorating". "It is infinitely more attractive than the gloomy aspect of passivity apparently offered by Darwinism" (Schrödinger 1967[1958], 107). But he soon corrects his own fancy take on Lamarck by declaring that "unhappily Lamarckism is untenable. The fundamental assumption on which it rests, namely, that acquired properties can be inherited, is wrong". Schrödinger here oscillates between his own *affective* liking of Lamarck and his *rational* awareness of the limitations of Lamarck's ideas and resolves this conflict between his emotions and reason with an attempt to propose a correction on Darwinian "mechanism of selection" in a way that would still make it possible to inherit the acquired characters (Schrödinger 1967[1958], 107–114). Schrödinger's Darwinism here is rather shoddy and incorrect and even when he declares that Lamarckism is untenable, his own loyalty to Lamarck is undeniable. And that is because Schrödinger thinks that

> our biological future, being nothing else but history on a large scale, must not be taken to be an unalterable destiny that is decided in advance by any Law of Nature. . . . We are, I believe, at the moment in grave danger of missing the "path of perfection".
>
> (Schrödinger 1967[1958], 115)

By the time Schrödinger gave these lectures in Dublin in 1956, Richard Semon had long committed suicide (in 1918) and his work on *Mneme* was widely discredited on the ground of its affinity for the Lamarckian idea of 'inheritance of the acquired characters'. For the biography of Richard Semon and a critical review of his work, see (Schacter 2001). Schrödinger does not discuss any of these criticisms of either Semon or Lamarck.

The similarity of the semantics

After the Indian mysticism, it is Schrödinger's engagement with Semon and Lamarck that has made historians of *What Is Life?* uncomfortable. There indeed are not clear evidences, as historian Yoxen thinks, to suggest that the concept of the code-script could be traced to the idea of engram in Semon's work on *mneme*. However, what is most interesting is the similarity of language and semantics in all four texts discussed here – *My View of the World*, the first draft written in 1925 and the last in 1961; *What Is Life?* written in 1944; and *Mind and Matter* written in 1958. Schrödinger writes in 1924–25/1961, "each of these bodies are at the same time blueprint, builder, and material for the next one, so that a part of it grew into a copy of itself" (Schrödinger 1964, 26). And, Schrödinger writes in 1944, "the chromosome structures are . . . instrumental in bringing about the development [of the offspring] . . . they are architect's plan and builder's craft – in one" (Schrödinger 1967[1944], 22). Further, in 1958 discussing how the enormous evolution of the past is brought about by innumerable individual chisel strokes, Schrödinger writes, "for we ourselves are chisel and statute, conquerors and conquered at the same time" (Schrödinger 1967[1958], 100). If we look at the similarity in semantics – 1) "Each body is at the same time blueprint and material for the copy" in 1925/1961, 2) "chromosome structures are architect's plan and builder's craft at the same time" in 1944, 3) and "we are chisel and statue at the same time" was written in 1958. This similarity is not just a matter of semantics; there is an underlying continuity in thought and a compelling search for answers to questions held close to his heart – the questions on inheritance of spiritual and somatic memory explored in *My View of the World*, on genetic inheritance debated in *What Is Life?*, and on inheritance as path to perfection for human evolution expressed in *Mind and Matter*. Synchronically speaking, whatever may be the connection of the gene as code-script with other scientific disciplines, diachronically placing *What Is Life?* in Schrödinger's life compellingly shows that the semantics of the concept of the gene as repetitive (inherited) and still varying unit of copy (each body is the blue print and material for the copy) was imagined when the discipline of cybernetics was not even born.

Conclusion

> Call to mind that sense of misgiving, that cold clutch of dreary emptiness which comes over everybody, I expect, when they first encounter the description given by Kirchhoff and Mach of the task of physics (or of science generally): "a description of the facts, with the maximum completeness and maximum economy of thought".
>
> (Schrödinger 1964, 3)

Many philosophers since Alexandre Koyre have argued how science cannot advance or acquire meaning without powerful assumptions about the world, without prior metaphysics. Several examples of the fruitful influence of such prior metaphysical and philosophical assumptions about the world on science have been now widely debated (Dupre 1993). Schrödinger himself begins his little book *My View of the World* with an argument how metaphysics does not form part of the house of knowledge but it is the scaffolding, without which further construction is impossible. In fact, as just quoted, Schrödinger even feels a "dreary sense of emptiness" thinking and doing science as mere description of facts.

Contrary to his own wishes, historiography of *What Is Life?* has consistently ignored Schrödinger's metaphysics, his emotional and existential *a priori* – the compelling search for the unity of existence – that remained the backbone of his thought throughout his life. Clearly, historians found his belief in the Eastern mysticism and his consistent references to Semon and Lamarck uncomfortable to be included in the discussions on the influence of *What Is Life?*. In ignoring Schrödinger's metaphysics, as I have discussed, the historiography of *What Is Life?* has strictly reviewed the influence of the book for its thesis of reductionism only and thereby made "the problem of reduction in biology far less daunting" as Olby puts it (Olby 1971, 125). Thereby, not only scientists so impressed with *What Is Life?* but even historians have remained wedded to the image of molecular biology as reductionist – what did not fit in this paradigm was unimportant, discarded, relegated to biography, or treated like aberration, appendage, or hobby. This way, not only that *What Is Life?* is only synchronically valued in the history of molecular biology, but that such history is presented in a progressivist fashion – only those contributions count that add up to the "advancement" of science. Such history of modern science validates only that metaphysics that assumes the material universe is fully law-governed, deterministic, and thereby a fully intelligible structure. Any rejection of such metaphysical assumptions is considered as "disorder of things" (Dupre 1993, 2). Such history of science is committed to the concept of science as nothing but "radical empiricism". Not only that science here happens "without knowing subject", it takes place as formulation of "disembodied ideas" (Amsterdamski 1975, 51; Söderqvist 1996), but also that the scientist-subject is imagined as a neo-Kantian ideal. He is first and foremost only a rational individual making theory choices based purely on empirical evidences. His science is thus detached from the rest of his emotions, affects, metaphysics, world views, existential desires, and fears and phobias; in short, the rest of his thought and his life experience (Söderqvist 1996, 2007).

For the whole of his life Schrödinger profoundly reflected on the nature of consciousness and continuity of somatic and spiritual memory across generations and the resulting oneness and eternity of the Self. Placing

What Is Life? diachronically in Schrödinger's life and thought, I wish to claim that the stark physicalism and reductionism of *What Is Life?* emerge from the metaphysical search for the permanence and stability of genetic material contributing to the immortal "I" so presented in the epilogue. His emphatic empirical declaration that the gene is nothing but physical and chemical entity, therefore, emerged from an entirely non-empirical and non-reductionist place. In a way, therefore, considering Schrödinger's generative role, it could be argued that reductionism of structural molecular biology is rooted in the form of thought that the same science declares as irrational, discredited, and untenable.

References

Amsterdamski, Stefan. 1975. *Between Experience and Metaphysics: Philosophical Problem of the Evolution of Science.* Dordrecht-Holland: D. Reidel Publishing Company.

Ceccarelli, Leah. 2001. *The Cases of Dobzhansky, Schrödinger, and Wilson.* Chicago: University of Chicago Press.

Domondon, Andrew. 2006. "Bringing Physics to Bear on the Phenomenon of Life: The Divergent Positions of Bohr, Delbrück and Schrödinger." *Studies in History and Philosophy of Biological and Biomedical Sciences* 37:433–458.

Doyle, Richard. 1997. *On Beyond Living: Rhetorical Transformations of the Life Sciences.* Stanford: Stanford University Press.

Dupre, John. 1993. *The Disorder of Things: Metaphysical Foundations of the Disunity of Science.* Cambridge, MA: Harvard University Press.

Fischer, Ernst Peter. 1984. "Of One Single Being: An Introduction to Erwin Schrödinger." *Social Research* 58 (1):809–855.

Fleming, Donald. 1968. "Emigre Physicists and Biological Revolution." In *Perspectives in American History*, 152–189. Cambridge: Harvard University Press.

Judson, Horace. 1979. *The Eighth Day of Creation: Makers of the Revolution in Biology.* New York: Simon & Schuster.

Kay, Lily. 2000. *Who Wrote the Book of Life? A History of the Genetic Code.* Stanford: Stanford University Press.

Keller, Evelyn Fox. 1990. "Physics and Emergence of Molecular Biology: A History of Cognitive and Political Synergy." *Journal of the History of Biology* 23 (3):389–409.

Keller, Evelyn Fox. 2002. *Making Sense of Life: Explaining Biological Development with Models, Metaphors, and Machines.* Cambridge, MA: Harvard University Press.

Moore, Walter. 1989. *Schrödinger: Life and Thought.* Cambridge: Cambridge University Press.

Olby, Robert. 1971. "Schrödinger's Problem: What Is Life?" *Journal of the History of Biology* 4 (1):119–148.

Olby, Robert. 1994. *The Path to the Double Helix: The Discovery of DNA.* New York: Dover Publications, Inc.

Schacter, Daniel. 2001. *Forgotten Ideas, Neglected Pioneers: Richard Semon and the Story of Memory.* London: Routledge.

Schrödinger, Erwin. 1964. *My View of the World.* Translated by Cecily Hastings. Cambridge: Cambridge University Press.

Schrödinger, Erwin. 1967. *What Is Life? The Physical Aspects of the Living Cell, with Mind and Matter and Autobiographical Sketches.* Cambridge: Cambridge University Press.

Söderqvist, Thomas. 1996. "Existential Projects and Existential Choice in Science: Science Biography as an Edifying Genre." In *Telling Lives in Science: Essays in Scientific Biography*, edited by Michael Shortland and Richard Yeo, 45–84. New York: Cambridge University Press.

Söderqvist, Thomas, ed. 2007. *The History and Poetics of Scientific Biography.* Burlington: Ashgate.

Stent, Gunther. 1966. "Waiting for the Paradox." In *Phage and the Origins of Molecular Biology*, edited by John Cairns, Gunther Stent and James Watson. New York: Cold Spring Harbor.

Symonds, Neville. 1986. "*What Is Life?* Schrödinger's Influence on Biology." *The Quarterly Review of Biology* 61 (2):221–226.

Teich, Makulas. 1975. "A Single Path to the Double Helix." *History of Science* 13:264–283.

Yoxen, E. J. 1979. "Where Does Schrödinger's *What Is Life?* Belong in the History of Molecular Biology?" *History of Science* 17:17–52.

5

THE MYTH AND TRUTH OF BARBARA McCLINTOCK[1]

She imagined herself intelligent, rational, civilized, believing in intellectual progress, and the experiments revealed her to herself as timid, desperately anxious about the effect she was having on other people, full of shy equivocations and tricks, hysterically violent, and irrational in her judgments when her self was threatened, and at a deeper level terrified, terrified of the future and of the point of existence, of her instinctive self, and of God as the avenger and punisher.

– W. H. Auden describing the psychoanalyst Marion Milner's own portrayal of herself so written in her experimental diaries eventually published under the name Joanna Field in 1934[2]

Declining the offer of writer Arnold Zweig to be his biographer, Freud called the request a "threat" and wrote to the writer that he was "alarmed by" it. He further wrote, "Anyone who writes a biography is committed to lies, concealments, hypocrisy, flattery and even to hiding his own lack of understanding, for biographical truth does not exist."[3] Even when it may not be as severe as Freud's misgivings of biography as a lie, writing a life by way of (auto)biography is nevertheless a craft. The scientific biography poses an even stronger challenge because it has another purpose or level – biography serves as a lens through which science in a particular period can be viewed. So while, on the one hand, narrating a life inevitably creates a myth, on the other hand, narrating scientific life necessarily alludes to historical truth about science.[4] At a more basic level, however, these contradictions of scientific biography could also be interpreted to raise questions such as: Who is the scientist-subject doing science? What is science and how is it done?

In this chapter I want to engage with two biographies, one by Evelyn Fox Keller and the second by Nathaniel Comfort, of the Nobel Prize–winning molecular biologist Barbara McClintock to revisit the myth and truth about her life and science (Comfort 2001; Keller 1983). Although both biographies are relatively old by now, and although they have already been reviewed sufficiently, my intention in this chapter is to explore the relationship between the two contradictory levels outlined earlier – the scientist's subjectivity and its relationship with the making of objective science.[5] In doing this, I will use resources referred to in both biographies and not engage in original research to create a fresh look at McClintock's personality.

According to the much-debated storyline, for the first biographer Evelyn Fox Keller, McClintock was an exceptional scientist devoted to a holistic conception of life, whose science of maize genetics was intuitive, sensitive, dynamic, interactive, and flexible, inspired by her "feeling for the organism" (Keller 1983, xiv). The key part of Keller's account of McClintock eventually made McClintock a feminist icon. In her biography, Keller discusses how McClintock felt that she was "not listened to" and that she was "ignored" and even "rejected" for a long time because she reported that "genetic changes were under the control of the organism" and were not controlling the organism (Keller 1983, 139–144). Her work on transposition and mobile genetic material posed a radical challenge to certain dogmas held by mainstream molecular biologists at that time. Such science firmly believed in genes as physical and chemical entities with fixed positions on chromosomes. It also believed in the one-gene one-enzyme thesis of the time that eventually contributed to the so-called central dogma of molecular biology: genetic information flows in one direction only, from DNA to RNA to protein (Crick 1970). Genes were thus controlling the organism and were not being controlled by the cytoplasmic (cellular) environment. Keller discusses in her biography how McClintock felt that she did not fit in such a scientific environment and how she felt she was "being ridiculed" and being told that she was "really mad" (Keller 1983, 140). As a result of such reaction, McClintock withdrew from her active presence in the scientific community in the 1950s and retreated to her laboratory where she worked in isolation for decades until in the 1970s when gene regulation was discovered by Franc̦ois Jacob and Jacques Monod. In the wake of this discovery, McClintock's work on transposition was rediscovered and eventually given due place in the history of genetic science when she was awarded the Nobel Prize in 1983. Keller's story of McClintock's life is an account of a woman scientist's conception of science and how her unorthodox views isolated her from the mainstream science (Keller 1983, xv). Many understood the story, though, as showing that women "see" scientific objects differently and how their science is holistic and hence radically different from the reductionism of male-dominated science.

McClintock's second biographer Nathaniel Comfort calls this story a myth. It is a myth not because it is wholly false but like any myth it contains both fact and fiction, Comfort clarifies (Comfort 2001, 4). In his detailed intellectual biography of McClintock, Comfort embarks on an energetic journey to separate "fact from fiction", "to peel back the mask" (Comfort 2001, 4). In doing so, Comfort organizes his own impeccable intellectual biography significantly around "dismantling the McClintock myth". It is important to point out that what Comfort calls "the McClintock myth" is not a unitary entity; it is composed of three separate parts – first, that McClintock was a loner; secondly, that everyone thought she was crazy; and thirdly, that she could "see" scientific objects differently. In Comfort's own words, "McClintock's private myth of being a loner whom everyone thought was crazy mutated into the public myth of McClintock as feminist idol" (Comfort 2001, 7). Comfort further argues that the parts of the McClintock myth emerged from McClintock herself in the first place. This myth was created in the 1970s when interest in her work was renewed and when historians and journalists requested interviews. According to Comfort, at that time she "settled into a pat version, easy to understand and ornamented with anecdotes and humour" (Comfort 2001, 4); this was "the version of her past that she could be comfortable with" (Comfort 2001, 18). For Comfort, this myth was further reincarnated, interpreted, and strengthened when Keller "attentively listened to McClintock's autobiography" and "took her at face value" (Comfort 2001, 5). Comfort alleges Keller's biography was based on "uncritical acceptance of McClintock's own private myth, the story she told about herself" (Comfort 2001, 4).

Inscribed in this tale of two biographies of myth-making and myth-breaking are different paradigms of the subjectivity of scientists. McClintock emerges in Comfort's biography as a brilliant, intimidating, and idiosyncratic but also methodical, rigorous, and gifted experimentalist. But in constructing this persona, Comfort considerably purifies McClintock of passions, emotions, and compelling affective drives, including her spirituality and her gender, as sources of her science. Comfort repeatedly shows in his intellectual biography that McClintock's theories on developmental control were not accepted by the scientific peers because her arguments were speculative and were not substantiated adequately by data (Comfort 2001, 129, 136, 140, 144, 156, 173, 184). But he barely pays any attention to the contradictions in the making of McClintock's affective self and the influence of these contradictions on the making of her science. Only in Keller's biography based on McClintock's own idea of her journey does she acquire a distinct subjectivity. But, in emphasizing the role of McClintock's inner psychic world – her fears, anxieties, aspirations, and ambitions – and the role of her gendered self in the making of her science, Keller downplays the role of intersubjective dynamics between McClintock and the scientific community. The difference between the two biographers is not entirely about evidences or about separating fact from fiction but about their adoption of two contrasting paradigms of a scientist's subjectivity: Keller's account foregrounds McClintock's affective self while Comfort foregrounds her rational self.

My own primary concern for this chapter is to understand the role of affects in the production of scientific knowledge. As Müller-Wille has pointed out in a review of Comfort's biography, the extent to which McClintock's academic style and her position was a matter of gender remains inconclusive in the biography (Müller-Wille 2002, 332). Also, as regards to the first and second components of what Comfort calls the McClintock myth emerging from McClintock herself, Comfort pays little attention, as Müller-Wille argues, to the question "why McClintock chose as she did" (Müller-Wille 2002, 332). Why did McClintock adopt such a private myth? In other words, the making of McClintock's affective self is not Comfort's concern. Instead of figuring out the extent to which the myth bears truth as Comfort does, I aim to ask a different question: How and why this private myth was in the making throughout McClintock's life and work? How this private myth was related to her self and her science?

My own claim is that this private myth emerged from the deep place in McClintock's affective and existential self, it was compellingly related to her objective science, her own subjectivity and her intersubjective recognition in science, and it was not something that she could have consciously "chosen" or "settled down for" while interacting with historians and journalists at the (convenient) later date as Comfort seems to suggest. Affect is here understood as subjective emotional experience that is from birth onwards co-constituted in ongoing relational systems. The theory of inter- subjectivity puts affect at the motivational centre of human psychological life (Stolorow 2011, 26). I aim to revisit both Comfort's claim of "truth" and his characterization of Keller's biography as "myth" by closely and comparatively reading both biographies towards my own reinterpretation of McClintock's work and life based on object-relational and intersubjectivity theories in psychoanalysis. In the background of this psychoanalytical perspective on McClintock's intrapsychic and intersubjective worlds that I wish to foreground is the question: How did McClintock's relationship with science's object and science's subject translate into a particular trajectory of her scientific contribution? How did McClintock's intrapsychic world of anxieties, fears, aspirations, desires, and ambitions manifest in her objective science? How did her intrapsychic world relate to the intersubjective world of science? But most importantly, how does this mutual recognition and dependence of the intrapsychic and intersubjective worlds relate to the historical legacy of reductionism in genetic science and in what way does it relate to the future development in molecular biology? This chapter, on the one hand, aims to provide an affective interpretation of the myth and truth of Barbara McClintock and her life and work by a close reading of two biographies, and on the other hand, it aspires to provide an alternative history of reductionism in molecular biology. It asks a question: Why and how in the microcosm of individual affective lives of scientists is it that the method and philosophy of reductionism was adopted, challenged, sustained, or questioned?

Capacity to be alone/incapacity for intimacy

In McClintock's life – reading through both biographies – two fundamental and contradictory affective trends emerge that had what I call a "non-responsive" relationship with both the object and the subject of science. On the one hand, she felt being rejected from the scientific community, and on the other, she showed a remarkable "capacity to be alone", as Keller calls it, quoting the psychoanalyst Donald Winnicott (Keller 1983, 15–37). Comfort alludes to a similar personality trait and names it an "incapacity for intimacy" (Comfort 2001, 31). These contrasting aspects of McClintock's psychic world – the wish to be recognized and the wish to be left alone – denote the fundamental conflict between the desire for recognition, and hence dependence on the other, and the declaration of omnipotence and hence independence of the self.[6] Both Keller and Comfort emphasize the fact that McClintock lived most of her life alone – physically and emotionally, for sure, but also intellectually. Keller thinks this was due to her aspiration for autonomy (Keller 1983, 15–37), and for Comfort it is the quest for freedom and intense need for solitude that was the key to her personal and intellectual life (Comfort 2001, 19). Both Keller and Comfort attribute self-empowering connotations to this personality trait (Keller 1983, 17). However, neither of them engages in more detail with the psychosocial or psycho-developmental origin of such capacity to be alone or incapacity for intimacy. In fact, engaging closely with Winnicott's theory in relation to McClintock's psycho-developmental history actually implies a different meaning of "capacity to be alone" than the one that Keller and Comfort propose.

Winnicott discusses the capacity to be alone in the following terms:

> Although many types of experiences go to the establishment of the capacity to be alone, there is one that is basic, and without a

sufficiency of it the capacity to be alone does not come about; this experience is that of being alone, as an infant and small child, in the presence of mother. Thus the basis of the capacity to be alone is a paradox; it is the experience of being alone while someone else is present.

(Winnicott 1958, 416)

Winnicott further relates the capacity to be alone to emotional maturity and posits that the presence of such

> maturity and the capacity to be alone implies that the individual has had the chance through good-enough mothering to build up a belief in benign and responsive environment. This belief is built up through a repetition of satisfactory instinctual gratification.

(Winnicott 1958, 416)

Winnicott argues that in someone's reliable presence without making demands – or in other words, in the presence of an ego-supportive environment that balances the ego-immaturity of the infant – the infant discovers and experiences personal "id impulses"[7] while being alone. And it is the integration of a multitude of such spontaneous and creative id impulses into ego-organization that "forms the basis for a life that has reality in it instead of futility" (Winnicott 1958, 419). Therefore, for Winnicott the "mature capacity to be alone" in adult life develops from the introjection (internalization) of an ego-supportive environment provided by the presence of good-enough mothering in infancy.[8]

Winnicott, however, asks in the same article if there is a way to speak of an "unsophisticated" or "pathological" form of being alone. He tries to answer this question in his work on the "false" and "true" self. The absence or failure of a good-enough mother that creates the ego-supportive environment at the pre-verbal developmental stage when the infant's own ego is weak produces "annihilation of the infant's self", which results in a radical interruption of the baby's sense of being and security (Winnicott 1958, 418). Winnicott argues that the self that results because of "inappropriate" responses from the mother is a "false self" which is built out of reactions from the outside or external stimuli, rather than emerging from the integration of one's own creative impulses, that is, id impulses into ego-organization. In the wake of a mother's failure to provide "adaptive care", especially ego support, the baby survives by means of the mind, whereby thinking becomes a substitute for maternal care and "the baby mothers himself by understanding, understanding too much" (Winnicott 1965, 156). Intelligence thus substitutes but also hides deprivation. In the absence of the integration of the spontaneous and creative impulses the false self is not only marked by a split – by "dissociation between intellectual activity and psychosomatic existence" – but is also plagued by feelings of unreality, deadness, futility, rigidity (Winnicott 1965, 143).

Following from similar early experiences, what Winnicott calls inability to make and find pleasure in relations with others (Winnicott 1965, 144) in the adult life comes remarkably close to what Comfort described as "incapacity for intimacy". Comfort highlights many examples of such personality traits in McClintock's life: she did not feel part of her family; she felt pain in having to conform to, or even function within, a group; she felt the social relations were not anchors but tethers; she never had any sexual or intimate relationships throughout her life (Comfort 2001, 20, 22, 25, 28). Winnicott's false self perceives dependence on others as a threat to its separate existence, and it is prone to severe or deny ties with others. In Winnicott's theorization the isolation or aloneness that may result from the prevalence of the false self is radically different from the mature capacity to be alone (in the presence of someone). The former fears and avoids ties with others and the latter happens (theoretically) in the presence of someone and is a marker of emotional maturity. In my argument McClintock's choice of aloneness was more a consequence of the former than the latter.

Keller equates McClintock's capacity to be alone with Winnicott's idea of the mature capacity to be alone, however, her description of McClintock's childhood relationship with her mother hints at the development of Winnicott's false self. According to the account of McClintock's elder sister, their mother was in acute stress when Barbara, the third child, was born, also because both parents had wanted a boy. Keller writes that the relation between the child and the mother was marked by tension from birth. In McClintock's own words, she was left alone with a pillow or a toy while she was as young as four months old (Keller 1983, 20). McClintock's own feeling deriving from the family legend is that in response to being left alone she "didn't cry, didn't call for anything" but as she grew older the tension with the mother kept on escalating to the extent that a year after the birth of the fourth child, a boy, Barbara, at age two and a half, was sent to live with her paternal uncle and aunt where she had good care and had a good time and did not miss the home (Keller 1983, 20). On her return, however, her relationship with her mother did not improve and instead became more distant. From the age of three she did not allow her mother to touch her (Comfort 2001, 20).[9] McClintock herself thought that the tension with her mother was responsible for her self-sufficiency, and her growing up as a solitary and independent child (Keller 1983, 22).

Since early childhood McClintock's personality structure shows the signs of a split between mind and psychosomatic existence in which, in Winnicott's terms, the "intellectual side was overdeveloped" – the core characteristic of Winnicott's false self. As a child she was not only an avid reader but she loved to sit alone and "think about things" (Keller 1983, 22). The adolescent Barbara found tremendous joy in "solving puzzles", and since her childhood she had also displayed a "striking capacity for autonomy, self-determination and total absorption" (Keller 1983, 26). Her preference for the mental existence was so profound that the body was a "nuisance" to her; she wished to be "free of the body"; "the body was something you dragged around" (Keller 1983, 36). McClintock's persona was shaped by a clear split between mental and psychosomatic existence – the core characteristic of what Winnicott calls the "false self". In addition to the split between mind and bodily existence, in Winnicott, the false self (that has emerged in the absence of the reliable primary care and hence suffered from the fear of annihilation early on) loses faith in the benign external world and retains persecutory anxiety throughout adult life.

Winnicott's false self, however, means that the brilliant intellectual self that did the pioneering work in transposition and movable genetic material and the self persistently afraid of and doubting the other, the self that consistently felt rejected, laughed at, not valued by the scientific community, and also the self whose scientific objects did not respond to her theoretical aspirations and ambitions, were inseparable parts of the same subjectivity. McClintock's lifelong feeling that she was rejected by her peers was not a myth. It was not something that she could have adopted later in her life as a convenient and comfortable story to be presented in front of media, as Comfort seems to think. It came from a deeply and compellingly affective place shaped by her developmental socio-psycho-somatic history. The pertinent issue to probe is how McClintock's intrapsychic world shaped her approach to her science's objects and subjects.

McClintock's intrapsychic world specific to her developmental psychoanalytical history – her desire for omnipotence than contact, her persecutory anxieties, her split personality overdeveloped for the preference for the mental existence at the cost of the psychosomatic one – shaped her approach to her science's object and subject. Her gendered subjectivity has to be understood in this light of her developmental history. In the next section I discuss the affective shaping of her objective science and how her objects of scientific investigation became progressively unresponsive to her theoretical and philosophical aspirations and ambitions. I show how McClintock's intrapsychic world was in complementary relationship with the intersubjective scientific world in a way that the dyad of the self and the other fail to recognize each other and how in this sense McClintock's objective science could be described in terms of Ian Hacking's "super-duper inter-subjectivity" (Hacking 2012, 20). In the final section I argue how this non-recognition point at the absence of what Jessica Benjamin calls the "third" (Benjamin 2006) as the dialogic space and how this absence hint at the incommensurability between McClintock's philosophical aspirations and the dominance of the prevailing reductionist dogma of the time. In conclusion, I posit that the McClintock myth has to be understood not as a combination of fact and fiction but as a product both of McClintock's own affective subjectivity and the intersubjectivity shaped by the historical legacy of molecular biology at that time.

Objectivity as intersubjectivity

Keller related the scientific community's rejection of McClintock's argument of controlling elements to her gender, whereas Comfort showed how McClintock's claims to controlling elements was rejected because it was not good and hence objective science. For Keller, it was a matter of contested subjectivity and for Comfort it was a matter of science's objectivity. The purpose of this chapter is to argue how both are related. Ian Hacking explains how Kant described objective/subjective in the *Critique of Practical Reason*:

> Practical principles [. . .] are subjective, or maxims, when the condition is regarded by the subject as valid only for his own will. They are objective, or practical, when they are recognised as objective, i.e., as valid for the will of every rational being.
>
> (Hacking 2012, 20)

In this sense, Hacking further explains how "objective" in Kant's *Critique of Practical Reason* means "inter-subjective, or rather, super-duper obligatory inter-subjectivity" (Hacking 2012, 20). It is clear in both Keller and Comfort's biographies what objectivity means in McClintock's science, but the idea of intersubjectivity is hardly explored. What would it mean to go beyond the gender as the only variant of subjectivity that both biographers explore and give an alternative view of "objective as super-duper inter-subjectivity"?

I want to take a brief detour to the concept of intersubjectivity in psychoanalysis before discussing objectivity as intersubjectivity in McClintock's science. Robert Stolorow broadly uses intersubjective to refer to any "psychological field formed by interacting worlds of experience" (Stolorow 2011, 23). Jessica Benjamin discusses intersubjectivity in Hegelian terms as mutual recognition of the self and the other with separate centre of experiences. As she explains, the individual not only grows in and through the relationship to other subjects but that humans have a fundamental need to be recognized and also the capacity for mutual recognition (Benjamin 1988, 19). In other words, in the words of developmental psychoanalyst Donald Winnicott, "I am seen therefore I am" is what creates a sense of being alive and real. Feeling real is more than existing. That the self exists only latently until it is subjectively realized within the responses of another person is a core part of Winnicott's theory of the self (Wright 2000, 91). Winnicott goes to the extent of saying that the need to find recognition for the self's experience is as basic as the need for satisfaction of bodily needs. But, the need for recognition gives rise to a paradox: recognition as the response from the other makes the feelings, intentions, and actions of the self meaningful, it allows the self to realize its agency, however, such recognition can come only from the other whom the self recognizes as a separate person. For Hegel the core of this problem is centred on the struggle between independence and dependence, between the self's wish for absolute independence and the self's need for recognition and hence the dependence on the other. In the words of Benjamin, "at the very moment of realising our own independence, we are dependent upon other to recognise it" (Benjamin 1988, 32). In the language of psychoanalysis this is the tension or vacillation between omnipotence and contact which Benjamin denotes to the "inherently problematic and conflictual make-up of the psyche" (Benjamin 1988, 30). Omnipotence is described as "a subjective state – a sense of complete control or influence – that the individual tries to bring about through action and/or fantasy" (Almond 1997, 3). Analytically speaking, this conflict translates into intrapsychic and intersubjective as complementary[10] ways to understand the psyche in relation. In other words, to recognize the intersubjective self is to realize the importance of the intrapsychic – the inner world of fantasy, anxiety, desire, and defence.

In the background of this discussion, it is pertinent to point out that both biographers have asymmetrically emphasized the intrapsychic and intersubjective aspects of McClintock's world. Keller unravelled McClintock's inner world – her anxiety, aspirations, ambitions, and disappointments – and related them with her science, but in doing so ignored or accepted on the face value the intersubjective dimension of McClintock's psyche and science. In contrast, Comfort discussed the dynamics of intersubjectivity and objectivity in McClintock's science but repudiated some of McClintock's intrapsychic perceptions as incorrect, myth, and fiction in the process. Later in the chapter I want to make an attempt to bring the worlds of intrapsychic and intersubjective together in the making of McClintock's science.

In the McClintock myth – the story about her self that McClintock told to herself – are the images of unresponsive environment, objective and subjective, from early on in her career. Here, I am less concerned to find out if the world was indeed as non-responsive as McClintock perceived and want to pay more attention to the way this perception emerged from McClintock's affective and existential subjectivity and the way these perceptions shaped her subjective response to her objective science. I first aim to discuss how in the early part of her career the scientific objects responded to her creativity, but her relationship with her scientific objects became progressively non-responsive which took a dramatic turn in her work on transpositions and controlling elements. At the same time, McClintock's relationship with science's subjects – the peer community – throughout her career remained distant, detached, often mutually critical, and in the later part of her career even outright rejecting.

McClintock's main focus since her graduation remained integration of cytology with genetics, i.e., cytogenetics, or in other words the study of cellular basis of genetic patterns. This was particularly significant in the times when genetic studies were increasingly being distanced from cytology, embryology, and from development of an organism. Historian of molecular biology Horace Freeland Judson describes the particular advance in genetic studies, quoting Sidney Brenner, as "Morgan's deviation". Brenner explains to Judson how Thomas Hunt Morgan, one of the pioneering scientists of genetic studies, originally started working on genetics because he was interested in development and embryology. But he gave up development and embryology because the problems were "intractable" (Judson, loc 4829). And even when Morgan went into the new field of genetics with a hope that it would cast light on development, all he and his students were able to do was genetics only. Morgan and his students worked on the common fruit fly *Drosophila*. Morgan discovered the mechanism by which the sex was determined in *Drosophila* and when one of his students, Alfred Sturtevant, developed a statistical method to map the relative distance of genes on a chromosome by mapping the linkage of many groups of genes on a chromosome (linkage groups were genes located closer on a chromosome and hence understood as inherited together). With the mapping of genes on a chromosome, the hereditary material became a string of beads, a line of points, palpably reduced to physical and chemical units, removed from their developmental role in space and time and still hailed as controlling the organism (Kohler 1994). Morgan's deviation opened up the reductionist program of genetic studies and separated it from understanding development and embryology.

Throughout her career, McClintock not only never quite separated cytology from genetics in her work, but later while working on transposition or what she called controlling elements she ambitiously and obstinately tried to combine development and genetics. Keller elaborates this theme in the enigmatic words of "feeling for organism" (the title of her biography) as McClintock's holistic approach to her science (Keller 1983). But Comfort counters Keller's projection of McClintock being holistic by arguing that she was no different than the prevailing scientific community, she also employed X-ray methods to artificially produce mutations in maize plants which at times even created monster chromosomes, which makes her as destructive as other scientists (Comfort 2001, 72). Comfort, however, downplays McClintock's lifelong theoretical conviction and her ambition to integrate cytology with genetics, or in other words, development with heredity. So obstinate was her intent that even when her objects of investigation did not respond to her desired outcome, or in Morgan's words, when the connection between development and genetics remained intractable, she still remained committed to it. In the early part of her career, soon after graduation, this involved establishing a relationship between chromosomes and genetic systems, in other words, locating the markers of known genetic traits of maize on the individual chromosomes. For McClintock, this was chromosomal analysis which was intended to map out processes by which sequences of genes underwent variation during inheritance. Both Keller and Comfort mention that locating linkage groups (sequences of genes inherited together) on chromosomes was well under way at that time in *Drosophila* but was not tried in maize. However, it was not that easy for maize as for *Drosophila*. It was much easier to locate genetic mutations such as different colour of eye or wing in *Drosophila*, but in maize it involved understanding intricate patterns of variegation on kernels. In case of maize, it therefore involved integrating plant breeding, patterns of colour, and texture of maturing plants that could be seen with the naked eye with microscopy of chromosomes. McClintock confides in her interview with Keller how "they thought me a little mad for doing this" (Keller 1983, 45). As I read it, Keller discusses how McClintock thought she was considered mad for combining plant breeding with genetics (Keller 1983, 47, 48), however, Comfort cites Keller and other sources of interviews with McClintock and interprets that McClintock thought she was ostracized for proposing to find linkage groups on particular chromosomes, which Comfort thinks was not so new at that time and hence considers it as part of her private myth, more fiction than fact here (Comfort 2001, 53). Whether one believes in Keller or Comfort, either way, McClintock's perception of a hostile professional environment, her work not being recognized by the *other*, persists. And this was so in spite of the fact that a highly specialized group of cytogeneticists was forming at Cornell at that time. At least two scholars from this group – Marcus Rhoads and George Beadle – remained McClintock's lifelong friends and supporters (Keller 1983, 48). It is in this context of McClintock's ambition to combine various aspects of biology – first cytology with genetics and later development with genetics – that the incidents of non-responsive scientific objects and even subjects increased as her career progressed. What I mean by non-responsive object is the way the object did not match, did not fit with, or, did not empirically respond to her theoretical ambition and intent. In other words, the way her empirical science failed to produce data and evidences in response to her theoretical ambitions.

In the late 1920s, after her graduation, McClintock followed the work of H. J. Muller on X-ray-induced mutagenetics, which expedited the production of mutations upon which genetic research greatly depended, which otherwise had to depend upon the rare chance of spontaneous mutations. Her work on X-ray-induced mutations resulted into a series of discoveries such as the ring chromosome and later the breakage-fusion-bridge (BFB) cycle. Keller describes the technique of X-ray induced mutations. It involved irradiating pollen with known dominant genes with particular traits and crossing them with kernels of plants carrying recessive genes of the same traits, or in other words, it involved crossing normal with damaged chromosomes (Keller 1983, 65). X-rays induced large-scale changes in chromosomal arrangements – translocations, inversions, and deletion of parts of chromosomes, which mutated into her important discoveries of BFB cycle and later in the most important work of her career on transposition qua controlling elements.

In the late 1930s McClintock's extensive experiments with X-rayed corn revealed a particular pattern which she later named as the breakage-fusion-bridge cycle. She observed that X-rays caused many chromosomes to break and to fuse again in normal or inverted sequence causing spontaneous and massive mutations. This breakage-fusion-bridge cycle continued within the lifetime of the plant until the broken ends were healed (Keller 1983, 81). For McClintock this breakage-fusion-bridging of chromosomes was not a random event caused by the X-ray damage but indicated specific patterns that controlled interactions among chromosomes while they also explained mutations. Following the behaviour of one of the monstrous chromosomes that had emerged from crossing over of the rearranged chromosome 9 with the normal chromosome 9 McClintock discovered patterns that led her to transpositions qua controlling elements.

Keller points out that the key to her work on transposition and her speculative theory on controlling elements, which was eventually rejected by the peer community, was the thinking that chromosomal change was a process (Keller 1983, 123) which, in my interpretation, involved time. Not only that this key thinking did not change throughout her life but that in adhering to this philosophical thinking in the absence of methodological and empirical backing that progressively McClintock's scientific objects became non-responsive and she herself withdrew from the professional world (Keller 1983, 123). This is not to suggest McClintock's scientific objects never responded to her. Working on breakage-fusion-bridge cycle involving X-rayed chromosomes, McClintock developed a powerful tool to produce directed mutations at a given location, the technique that eliminated the need for X-ray equipment (Comfort 2001, 83, 85). Such was her brilliance and her capacity for mastery over her method and technique. However, it was her work on mutable genes and transposition that created the speculative theoretical leap to a new theory of genetic control of development that made her scientific objects non-responsive. Among breakage, rearrangements, and deficiencies were transpositions – this was the term given to genes that changed position during detachment of a part of the chromosome arm and reattachment to a different chromosome by translocation or inversion (Comfort 2001, 91). In plants expressing such genetic instability she found evidences for regularly occurring and highly specific breaks in chromosomes, which she called dissociation (Keller 1983, 127). She argued that dissociation was actually controlled breakage; it was a response of one element of a chromosome in response to the signal sent out by another element. She called this system Ds-Ac system – dissociator and activator. Something in Ds triggers dissociation but something in Ac triggers Ds. These side by side sectors, one having a pattern reciprocal of the other, are what she called twin sectors. McClintock concluded that twin sectors accompanied most cases of chromosome loss during transposition (Comfort 2001, 93). She also argued that the adjacent (twin) sectors showed inverse relations in their mutation rates. She poetically described it as "one cell gained what the other cell lost" (Keller 1983, 123).

McClintock had thus recognized a pattern in transposition that she argued was not random and was not a result of the application of X-ray alone. But the discovery of transposition was not sufficient for her. As Comfort argues, her conceptual leap was to see that the pattern indicated control (Comfort 2001, 99). Control here meant to determine genetic action; this also meant that genes were not autonomous, independent entities. This conceptual speculation not only took her out of the genetics and brought her in the realm of development as Comfort shows, but it had a philosophical underpinning that subtly but most effectively guided the rest of her scientific journey, ultimately leading up to, on the one hand, the Nobel Prize, and on the other, her own self-imposed exile from the active scientific community. The most crucial and controversial aspect of her argument was the point that both timing and frequency of chromosome breaks were controlled (Comfort 2001, 105). The chromosome transposition as a process subjected to time and frequency was at the crux of her speculation of the theory of developmental control (Comfort 2001, 106). In other words, she speculated that the chromosomes controlled themselves while they activated right genes at the "right time" in the making of each cell and in this way exerted developmental control.

Morgan also similarly argued later in his career when he turned his attention back to embryology. He questioned the assumption that all the genes are acting all the time in the same way, which would not explain why some cells in the embryo develop in one way and some in another. He in fact brought the element of time in discussing the role of genes in developmental process by making an alternative view that different batteries of genes came into action as development proceeds (Keller 2000, 56). Morgan thus speculated that development implies that other agents on a chromosome call into action particular genes at the specific time and place. McClintock was struggling precisely on this issue of the role of the controlling elements in contributing to the development process by activating and deactivating genes at different times. McClintock also argued that there was a determining event in the development of a cell that decided its character because beyond which point it could not make certain kinds of tissues and played specific developmental roles from then on. This determining event was at that time studied by embryologists that incorporated an idea of time, which geneticists working on genes as hereditary units ignored.

As Steven Rose points out, geneticists' organisms were empty of time and internal content; there were only genes and phenotypes. There were no trajectories, no lifetime (Rose 1997, 113). Steven Rose explains how the concept of time, the idea of the direction of "time's arrow" is central to biology, how in physics time's arrow is reversible – the process can be repeated in both directions. But, living processes are irreproducible because the arrow of time irreversibly points in one direction only (Rose 1997, 14–15). The idea of the gene that emerged in the 1930s was not only reductionist – in the sense that it was nothing but a physical and chemical unit – but also mechanistic and static in that it did not include the concept of time. "Nothing in biology makes sense except in the light of evolution" are the famous words of population geneticist Theodosius Dobzhansky, which Rose interprets as "nothing in biology makes sense except in the light of history" – history of life on earth and history of an individual organism, its development through time (Rose 1997, 14).

Central in McClintock's idea of controlling elements was this idea of developmental time. As Comfort argues, this position on gene action subjected to time became her primary argument on controlling elements (Comfort 2001, 133). She continued working on transposition and discovered other major control systems in addition to Ds-Ac, for example, suppressor-mutator. Based on this work she speculated if there was a master switch that might have regulated a number of genes deciding when and to what extent to become active in the developmental trajectory (Comfort 2001, 133).

Although she eventually got the Nobel Prize for her discovery of transposition, it did not excite her. It was the role of transposition, its role in changing the pattern and hence exerting developmental control that remained her lifelong passion and conviction (Comfort 2001, 115). Even in the 1980s commenting on transposition being discovered in other organisms, she said, "Don't take that for granted. The real point is control. The real secret of all this is control. It is not transposition" (Comfort 2001, 115). The issue of the control of time and frequency of chromosome loss was always on her mind, even before she saw twin sectors (Comfort 2001, 102, 105, 106). This is the point I want to further elaborate before discussing how this underlying philosophical conviction led to the non-recognizing *other* – both object and subject – ultimately resulting into the feeling of being rejected herself by the scientific community.

McClintock's philosophical commitment to the controlling elements and hence to the central role of time in determining dynamics of gene action did not translate into method and did not result in empirical evidences. It is pertinent to note here that the gap I am talking about between her philosophy and method was the epistemological or experimental limitation of the historical time rather than her own failure. Based on the close reading of Comfort's intellectual biography I want to show that the gap always remained between her philosophical or theoretical ambitions and what the method could possibly deliver at that time and while she retained her faith in her philosophy throughout her life, her scientific object increasingly became empirically non-responsive to her ambitions owing to the lack of experimental method. This, however, does not reduce the importance of her philosophical convictions.

What I mean by the non-responsive object was visible in many of her, often bizarre, speculations and profound contradictions in her theorization. She first discovered that Ds and Ac were not that linked or located adjacent to each other as she had first thought. In some plants Ac was found far away from Ds, even located on a different chromosome. That means that the twin sectors were linked in some plants and not linked in others (Comfort 2001, 111). Another point she excitedly surmised first but soon became doubtful of was that Ac and Ds sometimes were not linked because the locus of Ac and even Ds had moved, they were movable elements (Comfort 2001, 113–114). She further theorized that these movable elements were neither geneticists' stationary chromosomal genes, nor, as it was surmised at that time, developmental biologists' cytoplasmic genes, but were "genelike particles" integrated on chromosomes and could even be found floating about the nucleus (Comfort 2001, 122). She argued that this regulation was achieved by chromosomal material called heterochromatin which did not contain any genes but was added to the genes for influencing gene action. The movable elements were tiny fragments of heterochromatin, which, when landed close to the gene inhibited its action and the inhibition was reversed when the movable elements moved again (Comfort 1999, 139). All through these – what now sound bizarre – speculations she projected that a chromosome was a complex, dynamic, and integrative structure. For her the idea of the gene as a unit of mutation and mutation as either chemical change (qua damage) or position effect were unacceptable (Comfort 2001, 129). Instead she proposed that mutations were chromosomal changes, changes in chromosome structure or chromosome elements other than genes (Comfort 2001, 133). All these speculations of alternative models to genes being autonomous and particulate matter were untenable by any experimental results and hence added in her reputation as crazy, maverick scientist. By 1956 she dropped the heterochromatin model, but still continued to argue that controlling elements were made of different stuff than genes (Comfort 1999, 150). The key to this conundrum was that for McClintock a gene was not a static unit but a sensitive dynamic structure which was subject to non-gene control (Comfort 1999, 142). This way she was even proposing to give up the term gene altogether. It is clear that McClintock did not side with the position of genes as strings of pearls, however, her own theory of genes remained vague, confused, contradictory, and shifting. What never changed, however, was her position that genes were controlled and that control subjected the organism to time-bound paths to growth and differentiation integral to the process of development (Comfort 1999, 151). The importance of control subjected to time was the crux of her unchanging and, what later started to sound unrealistic, improbable position.

She continued her struggle even on the face of her objects – concepts, entities, methods – not yielding to her ambitions. For years she remained busy explaining the mechanism for transposition because this, according to her, mediated genetics of development, but eventually she abandoned this idea and instead focused on the effects of controlling elements (Comfort 2001, 135; 1999, 149). Her ambition was to solve biology's big problems – development and evolution (Comfort 2001, 125), and she consistently projected control at the heart of all these problems even when she could not sufficiently, convincingly, and empirically demonstrate how. She even speculated on evolutionary change, the emergence of new life, as a loss of certain action in the cell, change in timing of such action, or change in locus of such action. She was even searching a single process at the heart of all the control (Comfort 2001, 124). Comfort names her ambition as "discipline-shaking theory", as "grand theorizing", as solving "fundamental problems" in her own words (Comfort 1999, 140–142).

Her position changed into an obstinate stand especially after her presentation at the 1951 symposium at Cold Spring Harbor, since when McClintock clearly felt not understood and rejected by the scientific community. These symposiums at Cold Spring Harbor were, as Comfort notes, premier professional meetings of geneticists. At that time, it was the location for the debate between the old radicals who argued against the gene and the young bacterial and viral geneticists who argued for it. And McClintock was firmly on the side of the old radicals (Comfort 2001, 11).

The discovery of transposition in bacteria in the 1960s, and eventually, once the full significance of transposition was recognized in the 1970s, McClintock's control argument was formally rejected (Comfort 1999, 137). And despite that she continued to believe in control even in the 1980s, even after she was awarded the Nobel Prize in 1992. Was she plain crazy? Was she just obstinate? Was she in denial?

Comfort argues that if the contesting point in McClintock's work was control and that if her arguments were not accepted and were challenged because she could not prove them, then gender discrimination as a category hardly helps to explain the reception of her work (Comfort 2001). Keller herself later meditated on the categories of sex, gender, science, and nature. She proposed to go beyond gender as the only variant of subjectivity and asks a question: "Is oppression ultimately the only important variable of gender?" (Keller 1987, 41). She also clarified that her notion of feminist science was not feminine science and that McClintock herself was not only not a feminist and throughout her life rejected such classification and considered science as a place where "the matter of gender drops away", but also that she refuted all stereotypic notions of femininity, thus complicating the role of gendered subjectivity in the making of her science (Keller 1987, 41). Keller also acknowledged that the contested reception of McClintock's work was more a case of the difference between science and non-science than the effect of gender difference on knowledge (Keller 1987). Other scholars have similarly questioned gender-specific discourse on method in science (Richards and Schuster 1989).

However, in this definition of what counts as good and hence objective science, science is imagined entirely as empirical process which is made of method and data and scientist is assumed as a rational entity making informed theory choices based purely on empirical evidences. For Comfort, McClintock's science was not good science because she could not find an appropriate method to empirically prove her conclusions. But science is much more than just empirical evidences. Many philosophers since Alexandre Koyre have argued how science cannot progress without powerful assumptions about the world, without prior metaphysics. Several examples of the fruitful influence of such prior metaphysical and philosophical assumptions about the world on science have been now widely accepted. However, as John Dupre argues these presuppositions must ultimately rest on evidences generated by empirical inquiry (Dupre 1993). McClintock's own challenge to the concept of the particulate gene and her insistence on the importance of timing of control in organism's development were philosophically anti-deterministic. But her failure to unify all of biology reflects, on the one hand, the underlying ontological complexity, which Dupre calls disorder of things (Dupre 1993, 7), and on the other, the unbridgeable gap between her philosophical assumptions and empirical findings. In spite of this failure, the importance of her philosophical assumptions that challenged the prevailing dominance of the deterministic and reductionist idea of the gene cannot be underestimated. Her philosophical stance on science was part of her larger world view. Comfort interviewed some of her close friends and colleagues who described her mysticism as the acknowledgement of a human inability to predict, know, and control things (Comfort 2001, 152). While she worked her whole life doing science towards prediction and control, her belief in such mysticism was not merely a paradox, it was a contradiction that fundamentally shaped who she was and what her science was. Keller focused on her portrayal of feminine method, and Comfort determined to break the myth of her being a feminist idol – both ignored that behind McClintock's science existed an underlying metaphysical world, there also existed the compellingly resilient affective self behind the method and myth.

Comfort conclusively declared that "emphasizing control makes her less of an icon, more of a human" (Comfort 1999, 138). But Comfort hardly engages with her humanity that made her so obstinately convinced about control. What remains to be explained is why she was able to adhere to such philosophical conviction despite the empirical obstacles and how this steadfast conviction despite rejection relates to her affective humanness. Keller locates these traits in her "idiosyncrasies of autobiography and personality" (Keller 1983, 97). The idiosyncrasy of her choice – she repudiated the empirical side and retained her metaphysics – needs to be explained.

As I have discussed in the previous section, what Keller calls idiosyncrasy can be explained in Winnicott's psychoanalytical terms of the "false self". In Hegelian terms the intrapsychic conflict between the preference for omnipotence qua independence from the other and need for the contact qua dependence on the other is fundamental to the human quest for recognition. As Winnicott's false self, McClintock's capacity to be alone as Keller calls it, or incapacity for intimacy as Comfort prefers to name it, emerges from this intrapsychic place of the preference for omnipotence rather than contact. This meant that it was not her gender but the lack of compelling need for the closeness to the other which gave her the enormous fortitude and courage to reject the dominant dogma of the time and remain obstinately committed to her own inner convictions on the face of the larger rejection. However, Winnicott's false self is plagued by persecutory anxieties at the same time. And while the intrapsychic preference for the omnipotence and hence the independence from the other is clearly the central aspect of McClintock's personality structure, the conflict between omnipotence and contact (between the desire for independence from and dependence on the other) is never quite completely resolved. McClintock's intrapsychic world was thus marked by this conflict and also by the persecutory anxieties that shaped her persistent feelings of rejection, the central dynamic of the intersubjective aspect of her science. This, however, does not explain her grand ambitions to do theory-shaking work, to do what was considered impossible at that time – to unify development, evolution, and molecular genetics.

Here I want to take a quick diversion to remark that the emergence and role of omnipotence in individual life acquires a slightly different meaning in psychoanalytical theory of *primary narcissism*. The retention of the sense of omnipotence and even grandiosity in adult life points to the insufficient assimilation of ego ideal into superego. Ego ideal refers to the blissful state of oneness and symbiotic union with the environment and the mother at the nascent stage of infancy, when all of an infant's needs are immediately satisfied and when the infant's idea of the self is not developed and differentiated from the environment and the mother. This blissful state is associated with the sense of wholeness and omnipotence, even grandiosity. The infant's eventual realization of the separateness of the self and the other causes the fundamental narcissistic injury, the ego's vulnerability is recognized, the illusion of omnipotence is shattered and the acute helplessness to satisfy instinctual urges on one's own is also experienced. The result is a humiliating sense of powerlessness (Chasseguet-Smirgel 1985, 44). What is called "object mastery" – the emergence of the ability to control one's environment and one's own self – ultimately compensates for this humiliation.

The sense of object mastery aspires to achieve ego ideal – it develops as a hope, guide, and promise that in growing up and in mastering the world the individual will recapture something of the lost paradise. When the goal of the ego ideal is omnipotence, grandiosity, and even rage, the super ego, on the other hand refers to socialization or normalization of individuals to cultural norms in adult life. The key issue is how in the individual's development into the adult life the state of ego ideal relates to the superego, how ego ideal of omnipotence and wholeness could be civilized and subsumed under superego. If ego ideal is too civilized, meaning subsumed too much under superego, it can hardly become the source of individuality and even non-identity and hence the source of resistance to less than perfect society. And if the ego ideal is not sufficiently matured by a certain degree of socially and culturally induced instinctual integration it retains the archaic forms of omnipotence and grandiosity (Chasseguet-Smirgel 1985, 188). The point is that a well-established superego in the place of ego ideal is not only not possible, but more importantly, it is not sufficient to heal the wounded narcissism. Chasseguet-Smirgel makes a profound statement, "man needs both bread and roses" (Chasseguet-Smirgel 1985, 187). The continuous striving for the "object mastery" – that is, seeking to influence the world and also seeking to perfect the self – are integral part of ego's striving for roses. I interpret McClintock's grandiose ambitions to do the impossible – unify all of biology – in other words, change the world and perfect the self was an expression of striving for the roses.

In McClintock's unfulfilled ambition, however, was also an impasse created between McClintock's intrapsychic response and the intersubjective world. This Jessica Benjamin calls complementarity, in other words, splitting relationships into push-me-pull-you, "doer-done to" dynamics found in all impasses. Benjamin further explains the complementary relations as the one in which the subject is unable to sustain the tension between the urge for independence and simultaneously seeking the recognition, in other words, the tension between omnipotence and contact. The breakdown that results from this inability to sustain the tension leads to a propensity to split relationships into what Benjamin calls "doer-done to" dynamics. In complementary relationships between the self and the other where each one feels right or the two perspectives are irreconcilable, such dynamics unfold in the form of – "either I am crazy or you are"; "if what you say is true, I must be very wrong, shamefully wrong, blind to what everyone else can see", "if you are strong, I must be weak" (Benjamin 2006, 123). The loss of agency of the self or the other is implied in complementary relationships. In the complementary structure "conflict cannot be processed, observed, held, mediated, or played with" while it pervades "helpless awareness of something that cannot be controlled" (Benjamin 2006, 122); such a relationship assumes the "form of victim/perpetrator and that both sides feel done to and do not feel themselves to be agents helping to shape a co-created reality" (Benjamin 2006, 123). Benjamin further argues that in complementarity there is an underlying symmetry in opposition of power relations: each feels unable to gain the other's recognition, each feels subject to the other's power, each feels his or her perspective is the right one. The essence of the complementary relations is the binary of resulting choices – submission or resistance to the other's demands.

Next I want to present some thoughts on the way McClintock's strongly held feelings of rejection from the scientific community point at the complementary relations of the dyad in which both McClintock and the peer group felt unable to co-create the place of the *third* – the co-created pattern of interaction and dialogue. My ultimate argument is that the absence of the third hints at the incommensurability of the philosophical assumptions on the concept of the gene entrenched on either side. It ultimately refers to the historical legacy of the way genetic science proceeded at that time and has implications for the future horizon of further progress rather than just a matter of contrasting subjective perspectives on objectivity of science that could only be comprehended in terms of declaring one side as fiction and other fact, as Comfort does.

I want to particularly engage with the 1951 symposium in which McClintock first presented her work on transposition and controlling elements since it is when she strongly felt being rejected. The dyad of complementarity, the doer-done to mode, came into existence in the mutual misrecognition of McClintock and her audience that more or less continued for the rest of McClintock's life. To place the symposium in the context of the mainstay debates in genetic science at that historical point – it took place in the years between Avery's demonstration confirmed by the Hershey and Chase experiment that the hereditary material was DNA and Watson and Crick's discovery of DNA as a double helix. The gene as particulate and chemically and physically a tractable entity arranged linearly on a chromosome was a widely accepted dogma of the time. At that time when Morgan even wrote that cytoplasm can be ignored genetically (Comfort 1999, 139).

Not only the complementarity of the dyad was expressed in McClintock's feelings of rejection, but it was also active on the other side. As Comfort argues, since Keller's biography, the scientists in question, those present in the 1951 symposium and other geneticists of the time, felt attacked by the feminist outsiders. They locked arms and insisted that McClintock was never ignored. Misunderstood, but not ignored! (Comfort 1999, 136). Comfort, reporting his own impression while extensively engaging with

McClintock's material – her field notes, lab notebooks, letters, handwritten thoughts, memoranda, draft reports (Comfort 2001, 12), and also quoting other scientists who were subsequently interviewed (Comfort 2011), describes her work as elliptical, difficult to understand. Comfort declares that the scientists present at the 1951 symposium "had been baffled by a brilliant but knotty concept, puzzlingly presented" (Comfort 1999, 136). These remarks clearly point out the complementarity of the dyad in the absence of the *third*, which is visible in Comfort's remarks such as "no formal discussion followed the talk, several printed comments on other papers mentioned her work" (Comfort 1999, 145) but with the comments such as "findings were important but interpretation did not quite fit" (Comfort 1999, 151). One scientist, even an otherwise admirer of McClintock, wrote, "I do not understand the detail of this work well enough to put my finger on what may be particularly significant" (Comfort 1999, 151). Comfort even wants to make a lot out of the fact that even when no discussion followed her presentation, in the photos taken after the session she was found having an animated discussion with the other participants – Comfort sounds as if the presence of such visual animation could be accepted as a surrogate for the real dialogue on her work (Comfort 1999, 145). Comfort ultimately pins down his argument to one place based on his own rigorous reading of McClintock's material and also quoting several of the participants. That is, McClintock's assertions about transposable genetic material as elements controlling development and eventually evolution could not be empirically ascertained, that she did not have enough data, that she was speculative, that scientists kept her control arguments at arm's length in the 1950s, and eventually she was not proven right once the full significance of transposition was recognized in the 1970s. Comfort finally concludes that McClintock's sweeping theory is now relegated to another idea in history of science which had once been plausible but is now seen as wrong. Comfort also mentions how McClintock was rather irritated for being told how her speculative theory was wrong (Comfort 1999, 152).

None of this would stand up as a candidate for a serious attempt to understand why McClintock remained so committed to the idea of control in the stark absence of the data. The degree of importance ascribed to data and evidences translates into a position that if she did not have data, "she was wrong" or "she was crazy", as one of the participants did describe her in a similarly complementary term, "she was either genius or crazy." While the reception of McClintock's work at the 1951 symposium may not be chauvinism of the male-dominated genetics community as some of the feminist scholars might have projected it based on Keller's biography, but dogmatism it nevertheless was. Those, except perhaps the controversial scientist

Goldschmidt, who engaged with her arguments on controlling elements did so only within the dominant paradigm of the Beadle and Tatum model of one gene one enzyme, which meant adhering to the reductionist concept of the particulate gene as a physical and chemical unit distinctly located on a chromosome unilaterally controlling an entire organism. Her attempt to present an alternative model of the gene that emphasized the pattern of behaviour of the entire genome with different levels of control was judged only as overly speculative, not testable, a little mystical, pretty cosmic, and was not seriously debated at all (Comfort 1999, 165).

It is not difficult to side with Comfort's challenge to the notion that McClintock's science was rejected because of gender discrimination; however, it is not easy to accept his suggestion that McClintock did not face any cold, silent, and rejecting treatment at the symposium. Comfort repeatedly justifies this treatment by almost suggesting that McClintock may not have deserved it but still she brought it upon herself by being speculative, for not having enough data and evidences. Comfort's assertions sound as if in the history of genetic science this was the first time ever that a scientist presented a speculative model or used metaphors and models that could not be tested and proven at that time. The most glaring examples of such speculative and rhetorical metaphors that had profound influence on the direction of genetic science are genetic information (Griffiths 2001), the gene as a code-script (Schrödinger 1944), or, not so positive example could be what eventually proved as utterly nonsensical world scripture models of George Gamow which was nevertheless taken seriously and debated.[11] None of these overly speculative models of the time met with such cold and silent treatment. My thesis is that none of these unproven models were rejected because they sided with the dominant dogma of the time, whereas McClintock's speculative model challenged such dogma.

I want to end this section by making two points. Firstly, McClintock's own feeling of rejection since her presentation in the 1951 symposium was not just a myth made partly of fact and partly fiction, something that she floated only retrospectively much after the symposium. McClintock's own feelings came from a deeply affective place that related to her own developmental history and her deep intrapsychic fears and anxieties; these were not the calculated constructions she would have settled down for at a later convenient date as Comfort suggests. McClintock shows personality structure of what Winnicott called "false self", which meant that she suffered from the persecutory anxiety whole her life, her life experience and her science consequently were intimately shaped by the non-responsive environment, both objects and subjects.

Secondly, McClintock's feelings about how the 1951 symposium participants responded and how her work was subsequently received point out the breakdown of the double direction of recognition, the absence of the co-created reality. Implied in this failure or breakdown is the incommensurability of the philosophical assumptions held by both sides and the larger historical legacy of the making of the reductionist science of the macromolecule of a gene. The truth and myth of McClintock's science must be evaluated in the context of how molecular biology has since evolved. Later in the chapter I discuss some of the latest in molecular biology to show how McClintock's philosophical convictions have eventually been received.

McClintock myth in the absence of the third: past legacy and future horizon of genetic science

At the time of the much-debated symposium of 1951, scientific discourses were predominantly preoccupied with explaining the structure and function of the gene. How to distinguish, individuate, and count genes on chromosome? Where does the gene begin and where does it end? How does it relate to a particular developmental effect or trait? These were the questions that had fundamentally driven molecular biology. Spanning over the first half of the twentieth century a series of discoveries postulated that genes were located at exact positions on chromosomes. With these discoveries, a series of highly influential dogmas emerged that postulated one-to-one functional correspondence between a gene and some developmental unit, these could be phenotypic traits – colour of eyes, hair, etc. – or an enzyme or a polypeptide – a protein molecule. At the time of the symposium in 1951, Beadle and Tatum's dictum on one gene one enzyme was the dominant dogma. Soon after the discovery of the double helix, the beauty and simplicity of the structure of the gene, its functional and structural correspondence with a phenotype or chemical trait, and accordingly the simple resolution of the riddle of life were widely celebrated. In this climate, McClintock's challenge to the reductionist physicalism of the gene as either position or chemical effect and her philosophical preference for a dynamic chromosome were not well received.

A decade later, these two poles of molecular biology – the one that began with Morgan's deviation from developmental biology, which opened up the highway to the reductionist concept of the gene, and the second, which rejected the gene as a physical and chemical unit in Goldschmidt and McClintock's work – were bridged. Jacques Monod's and François Jacob's discovery led to the distinction between structural genes and regulatory genes, which in turn led to the discovery of whole ensembles of genetic machinery of control. Structural genes code for the proteins and regulatory genes regulate the rate at which the structural genes are transcribed.

Regulatory genes are in turn structured by other structural and regulatory genes (Keller 2000, 55). Jacob and Monod called it the operon model, which referred to a cluster of regulatory and structural genes whose functioning is coordinated by the products of regulated genes located elsewhere in the genome (Keller 2000, 57). The difference between McClintock's and Jacob and Monod's regulatory model was the transposition of genetic material, i.e., none involved in the operon model. Still, for McClintock, Jacob and Monod's work was the vindication of her long-time emphasis on control and gene regulation and her insistence that such control was exercised in response to changes in cellular environment. Since Monod and Jacob's Nobel Prize–winning operon model, the work on regulatory elements in the genome has proliferated, and echoing McClintock's own scientific vocabulary, these are often called promoter and terminator sequences, or leader sequences, some others are called activator elements (Keller 2000, 58).

Furthermore, since the sequencing of the human genome it has become abundantly clear that only about 3% of the human genome codes for amino acids, meaning only 3% is composed of coding or structural genes and the rest is involved only in regulatory activities of some kind. Some of these genes provide only the template for the gene products, some others provide only the specific site at which coding happens. This means that the functional gene contains many more elements than just the coding sequence, as it was originally thought. Further research has made it apparent that a typical gene includes, in addition to structural or coding regions, regulatory regions called promoters and terminators, transcription machinery, enhancers, and exons and introns. The regulatory regions were described as junk DNA in the human genome sequencing, but in the last ten years it has become increasingly clear that the non-coding, regulatory regions have a profound effect on the timing and content of gene expression. Another caveat is that the control or regulatory regions, the non-coding genetic material, do not always have to be located close to the structural or coding sequences. They might even be located on a different chromosome or far away on the same chromosome and even be entirely mobile on a chromosome. More so, the coding regions called exons are not only fragmented but are interspersed among extensive non-coding regions called introns. That means that not even the coding or structural genes have structural unity. Where then should we count as the beginning and end of the physical or chemical unit called the gene located at the distinct position on a chromosome? Not only that the structural integrity of the functional gene – meaning that the gene is a particular DNA sequence located at the exact location on a chromosome – came under question, but it also has become clear that the gene regulation, that is, under what circumstances genes are switched on and off, involves interaction with many more genes and also cellular and environmental stimuli. Some scholars call this "causal democracy" (many cellular, genetic, and epigenetic processes are causally *equally* necessary in determining the developmental outcome). Furthermore, it has become most difficult to prove the idea that each developmental trait is explained by the action of a particular gene or genes. There could be a vast number of coding sequences that could code haemoglobin and insulin for example. Philosophers of molecular biology often explain this impossibility to relate molecular structure with biological function in terms of two concepts: *supervenience* and *multiply realize*. The concept of supervenience is from evolutionary biology, which is about natural selection. It means that from a range of available variations in nature, a particular effect or trait is chosen in natural selection; this effect is blind to structural differences in DNA. In other words, it means that thousands of combinations of DNA sequences could possibly code for one trait to come in effect. Natural selection in fact maintains such structural diversity. In other words, there cannot be a single molecular structure for a single functional trait. In terms of gene expression, it translates into what is called a "many-many" problem (a vast number of genes are responsible for a vast array of developmental and regulatory activities that ultimately result in a particular trait). In other words, all traits require the action of many genes and many genes contribute to the development of more than one trait. This means that it is impossible to find a distinct spatio-temporally unique, structurally locatable, and functionally identifiable entity called a gene that is responsible for one particular trait or expression. In the light of the findings of the current science, it has now become undeniable that genes determine very little, if anything, towards the developmental trait, and they are just one of many actors in an extraordinarily complex genetic and epigenetic scenario.[12]

More recently, the results of the sequel to the human genome project called ENCODE (Encyclopedia of DNA Elements) were released in 2012, radically redefining what is called a gene.[13] One of the important findings of the ENCODE project is that not only is 80% of the genome involved in controlling or regulating the expression of 1% of genome coding DNA, but, most importantly, regulatory regions are dispersed throughout the chromosome – there are regulatory rich forests and poor deserts and regulatory sites are located far away, even on another chromosome, from the functional or structural genes (Henikoff 2007). The working of the regulatory system is found to be so complex that it is no longer considered productive to fold it into the definition of the gene. In the recent past, the concept of the particulate gene has been revised to mean – "genes as subroutines in the hugely complex genomic operating system" (Gerstein et al. 2007, 671).[14] It was also discovered that the gene itself has a discontinuous structure – one gene could be continued inside another one's intron or one gene could overlap another without sharing exons or a regulatory system. At the end of the project various attempts were made to radically revise the definition of the gene. One such definition is pivoted not on structural elements but on functional products – "the gene is a union of genomic sequences encoding a coherent set of potentially overlapping functional products" (Gerstein et al. 2007, 677). Increasingly, it is more productive to define the gene in terms of what it is not than in terms of what it is. One such attempt defines the gene in these terms – "The gene is neither discrete . . . nor continuous . . . nor does it have a constant location . . . nor a clear cut function . . . and not even constant sequences . . . nor definite borderline" (Gerstein et al. 2007, 679). This means that it is impossible to designate one gene for one action or one-way flow of information from a gene to cell or even locate a gene on a chromosome as a composite of a series of nucleotides. By the beginning of the twenty-first century the concept of a gene as a discrete and spatio-temporally unique entity located on a chromosome has so much come under challenge that some philosophers and even scientists consider the gene as an obsolete idea.

This current science could be termed as the future horizon of the stony silences that followed McClintock's presentation in the 1951 symposium, which eventually have powerfully countered the prevailing dogmas of one-gene one-enzyme theory and only one-way flow of information between the gene and the cell. Even when McClintock could not really provide empirical evidence on her arguments on controlling elements, her broader philosophical assumptions on the role of timing in activating and deactivating genes and the genome being a dynamic and complex entity have been clearly vindicated. In the light of these latest developments in genetic science, McClintock's resilient adherence to her philosophical commitments, even on the face of the absence of empirical evidences, stony silences, and rejections, sounds like a super good science.

Conclusion

In conclusion, I want to highlight two points. Firstly, my reading of the two biographies of McClintock's life raises important questions about the relevance and usefulness of psychoanalytical theories for understanding scientific subjectivities. The current literature on the history of scientific subjectivity predominantly constructs the scientist-self that is the neo-Kantian ideal – a unified and wilful, self-determined, self-regulated, active and autonomous, rational subject wilfully driven by social and scientific ethos. This I have discussed at length in Chapter 2. The psychoanalytical approach makes it possible to understand the knowing subject as a funda-mentally feeling, suffering, heterogeneous, contradictory, and incoherent being – the subject that is fictional and real at the same time and comes into being in relation to others. I intend here to reinterpret two aspects of Comfort's claim about what he calls the McClintock myth – first, that the myth emerged from McClintock herself, that she "settled into" a pat version of her past that she was comfortable with; and second, that Keller uncritically accepted McClintock's own (autobiographical) private myth which mutated into a public myth of McClintock being a feminist icon. This is the myth that Comfort attempts to dismantle by separating fact from fiction. By using a developmental psychoanalytical approach, I show that what Comfort calls McClintock's private myth not only emerged from a deeply and compellingly affective place in McClintock's life, but it was also integral to and fundamentally formative of who she was, as a woman and a scientist. The psychoanalytical approach to looking at McClintock's developmental psycho-social history shows that the self that was responsi-ble for the brilliant work on transposition, the self that obstinately retained certain philosophical assumptions despite the lack of evidence, the self that did not feel at home with her own family, the self that did not have any intimate relationships throughout her life, and the self that suffered perse-cutory anxieties of rejection from the scientific peers were all part of the same affective subjectivity.

Jane Flax – a feminist scholar and a practicing psychoanalyst – describes how psychoanalysis "presents the best and most promising theories of how a self that is simultaneously embodied, social, 'fictional', and real comes to be, changes and persists over time" (Flax 1990, 16). Even when McClintock's feelings of rejection from her scientific peers might have been partly "fictional", the feeling of rejection was real for McClintock, it was not some part of a story she "settled into" at a convenient time in the 1970s. Moreover, this feeling of persecution was part of the same subjectiv-ity that underlay her capacity to be alone. The lack of compelling need for closeness to the other gave her enormous fortitude and courage to reject the dominant dogma of the time and remain obstinately committed to her own inner convictions on the face of the alienation from both science's subjects and objects.

Secondly, the chapter shows how McClintock's stories about the self and science that she told to herself not only emerged from the deeply and compellingly affective place but they also manifested at a particular his-torical juncture related to the legacy of reductionism in molecular biology.

The chapter is an attempt to illustrate the way the method and philosophy of reductionism was challenged and questioned in the microcosm of the affective and the cognitive world of McClintock's life and work. The chapter is therefore an alternative interpretation of the history and legacy of reductionism in molecular biology.

Notes

1 A version of this chapter was first published as Shah, E. 2016. "A Tale of Two Biographies: the Myth and Truth of Barbara McClintock." *History and Philosophy of Life Sciences*, 38 (18). https://doi.org/10.1007/s40656-016-0119-9.

2 As quoted in Auden (2000, 113).

3 As discussed and quoted in Phillips (2000, 31).

4 For further discussion on biographies as sources of the history of science, see Russell (1988), Selya (2003), Nye (2006), and Söderqvist (1996, 2007, 2011). For the discussion on scientific biographies as source of history in life sciences, see Abir-Am (1982, 1991).

5 A recent special volume has a few studies and a detailed review of the current literature on the way objectivity and modern scientific self are in part emotional construct (White 2009). For a discussion on feminist epistemologies and their critique of the scientist-subject, see Shah (2013).

6 Omnipotence is described as "a subjective state – a sense of complete control or influence – that the individual tries to bring about through action and/or fantasy" (Almond 1997, 3).

7 Id, ego, and superego are three aspects of the psychic apparatus defined by Sigmund Freud. According to this structural model of the psyche, id is the set of uncoordinated instinctual trends, the superego plays the critical and moralizing role, and the ego organizes the personality by mediating between the desires of the id and the restrictions of the superego.

8 In Winnicott's object relations theories there is an element of essentialism in two forms. Winnicott's infant develops into true self that is remarkably degendered and asexual. However, the most troublesome part of Winnicott's theories is essentialization of the role of good-enough mother. Despite the claim of some of the object relations theorists that the mother-child relationship is mutually constitutive and reciprocal, in Winnicott's theories especially the mother appears primarily as child's object. The mother disappears as a separate person, her subjectivity is not explicitly problematized, and the story of the pre-verbal development of human psycho-soma is told from the child's perspective. Object relations theories thus have failed to create a sustained critique of the role of gender in the making of the individual adult identity and also the role of gender in the understanding of good-enough mothering for such identity formation. Winnicott's object relations theories are especially ahistorical in this sense compared to Freud's superego which suggests that the child also internalizes the mother's, the father's, and other caretakers' past and present object relations. In this sense, to an extent, the parents' entire social histories become part of the child's self. Even on the face of these limitations, however, there is a great value in Winnicott's object relations theories and its crucial assumption that the subject comes to being in the field of the other,

that human beings by nature are "object seeking", that they need real and not merely projected or narcissistic relations with others. For further discussion on the question of gender in Winnicott's object relations theories, see Flax (1990, 108–132). And for the feminist correction on Winnicott's concept of good-enough mothering, see Benjamin (2006). In a series of publications Benjamin develops the idea of the third as the double directionality of recognition according to which the other, including in the dyad of mother and infant, is there to receive the communication that allows the other to become real and by that means opens the space of the thirdness to negotiate the differences and connect in the real way and not just projected form. For the further powerful critique of Foucauldian historicism against Lacanian psychoanalysis, see Copjec (1994). For the feminist interpretation of ahistoricism in psychoanalysis, see Minsky (1990).

9   I often found Comfort's attempts at myth-breaking unsympathetic to McClintock's views of her developmental past. Comfort reports that McClintock attributed her alienation from her mother to the fact that her mother did not like babies, including her. Comfort however quotes McClintock's elder sister to clarify that their mother, Mrs McClintock, was rather stressed out during this period due to the birth of the fourth child and due to the burden of raising all four children on the limited income of her husband. Both these versions of Mrs McClintock's disposition towards babies could have been possible at the same time – the mother not liking babies, including Barbara, and also that she was stressed out bearing the burden of caring for them. However, Comfort chooses to call McClintock's feeling that her mother did not like her as a baby a fiction and her sister's version of family history a fact (Comfort 2001, 20). This is an example of how in his biography Comfort sometimes unsympathetically discounts McClintock's feelings as a myth. Instead of taking issues with Comfort's such painstaking efforts at myth-breaking, what is important to note here is that there clearly was a sense of alienation between the mother and the infant daughter, whatever may be the reasons. And this sense of alienation foundationally shaped McClintock's life experience.

10 Jessica Benjamin, drawing from the Hegelian concept of recognition, defines complementarity as a state of the dyad in which the self and the other are unable to experience the two-way direction of effects. This is the state of breakdown of the mutual recognition in which both sides feel victim to the other's non-recognition (Benjamin 2006, 122).

11 For further discussion on the role of metaphor in the history of molecular biology, see Doyle (1997).

12 This discussion has referred to the following sources: Keller (2000, 55–60), Rosenberg (2006, 30–32, 119–125), Stotz and Griffiths (2004), Griffiths and Neumann-Held (1999), Griffiths and Stotz (2007), and Rose and Rose (2012).

13 ENCODE (Encyclopedia of DNA Elements, launched in 2003) was a combined work of 400 scientists from around the globe. The results were published in *Nature* with more than 30 companion papers published in other journals. The project was initially launched to determine the role of the junk, or non-protein coding, DNA, but it has contributed to the debates on the genetic variant causing disease, and has also given insights into human evolution. For the concluding publication of the ENCODE project consortium, see Consortium (2012). And for further online publication of the data, see Consortium (2011). For a

commentary on the completion of the pilot project in 2007, see Weinstock (2007). For a review of the ENCODE results, see Kellis et al. (2014).

14 For a further discussion on ENCODE, see Henikoff (2007). For a commentary on the completion of the pilot project in 2007, see Weinstock (2007).

References

Abir-Am, Pnina. 1982. "How Scientists View Their Heroes: Some Remarks on the Mechanisms of Myth Construction." *Journal of the History of Biology* 15 (2): 281–315.

Abir-Am, Pnina. 1991. "Nobelesse Oblige: Lives of Molecular Biologists." *Isis* 82 (2): 326–343.

Almond, R. 1997. "Omnipotence and Power." In *Omnipotent Fantasies and the Vulnerable Self*, edited by C. S. Ellman and J. Reppen, 1–37. Northvale, NJ: Aronson.

Auden, W. H. 2000. "To Unravel Unhappiness." In *Winnicott Studies Monograph Series: Art, Creativity, Living*, edited by Lesley Caldwell. London: Karnac Books.

Benjamin, Jessica. 1988. *The Bonds of Love: Psychoanalysis, Feminism, and the Problem of Domination*. New York: Pantheon Books.

Benjamin, Jessica. 2006. "Two-Way Streets: Recognition and Difference in the Intersubjective Third." *Difference: A Journal of Feminist Cultural Studies* 17 (1):116–146.

Chasseguet-Smirgel, Jenine. 1985. *The Ego Ideal: A Psychoanalytic Essay on the Malady of the Ideal*. Translated by Paul Barrows. New York: W.W. Norton & Company.

Comfort, Nathaniel. 1999. "'The Real Point Is Control': The Reception of Barbara McClintock's Controlling Elements." *Journal of the History of Biology* 32 (1): 133–162.

Comfort, Nathaniel. 2001. *The Tangled Field: Barbara McClintock's Search for the Patterns of Genetic Control*. Cambridge: Harvard University Press.

Comfort, Nathaniel. 2011. "When Your Sources Talk Back: Towards a Multimodal Approach to Scientific Biography." *Journal of the History of Biology* 44:651–669.

Consortium, The ENCODE Project. 2011. "A User's Guide to the Encyclopedia of DNA Elements (ENCODE)." *PLoS Biology* 9 (4):1–21.

Consortium, The ENCODE Project. 2012. "The Integrated Encyclopedia of DNA Elements in the Human Genome." *Nature* 489:57–74.

Copjec, Joan. 1994. *Read My Desire: Lacan against the Historicists*. Cambridge, MA: The MIT Press.

Crick, Francis. 1970. "Central Dogma of Molecular Biology." *Nature* 227:561–563.

Doyle, Richard. 1997. *On Beyond Living: Rhetorical Transformations of the Life Sciences*. Stanford, CA: Stanford University Press.

Dupre, John. 1993. *The Disorder of Things: Metaphysical Foundations of the Disunity of Science*. Cambridge, MA: Harvard University Press.

Flax, Jane. 1990. *Thinking Fragments: Psychoanalysis, Feminism, and Postmodernism in the Contemporary West*. Berkeley: University of California Press.

Gerstein, Mark, Can Bruce, Joel Rozowsky, Deyou Zheng, Jiang Du, Jan Korbel, Olof Emanuelsson, Zhengdong Zhang, Sherman Weissman, and Muichal Snyder.

2007. "What Is a Gene, Post-ENCODE? History and Updated Definition." *Genome Research* 17:669–689.

Griffiths, Paul. 2001. "Genetic Information: A Metaphor in Search of a Theory." *Philosophy of Science* 68 (3):394–412.

Griffiths, Paul, and Eva Neumann-Held. 1999. "The Many Faces of the Gene." *BioScience* 49 (8):656–662.

Griffiths, Paul, and Karola Stotz. 2007. "Gene." In *The Cambridge Companion to Philosophy of Biology*, edited by David Hull and Michael Ruse, 85–102. Cambridge: Cambridge University Press.

Hacking, Ian. 2012. "Objectivity in Historical Perspective." *Metascience* 11:11–39.

Henikoff, Steven. 2007. "ENCODE and Our Very Busy Genome." *Nature Genetics* 39 (7):817–818.

Judson, Horace Freeland. 1996. *The Eighth Day of Creation: Makers of the Revolution in Biology*. New York: Cold Spring Harbor Laboratory Press.

Keller, Evelyn Fox. 1983. *A Feeling for the Organism: The Life and Work of Barbara McClintock*. New York: W.H. Freeman and Company.

Keller, Evelyn Fox. 1987. "The Gender/Science System: Is Sex to Gender as Nature Is to Science?" *Hypatia* 2 (3):37–49.

Keller, Evelyn Fox. 2000. *The Century of the Gene*. Cambridge, MA: Harvard University Press.

Kellis, Manolis, Barbara Wold, Michael Snyder, Bradley Bernstein, Anshul Kundaje, Georgi Marinov, Lucas Ward, Ewan Birney, Gregory Crawford, Job Dekker, Ian Dunham, Laura Elnistski, Eric Green, Roderic Guigo, Tim Hubbard, Jim Kent, Jason Lieb, Richard Myers, Michael Pazin, Bing Ren, John Stamatoyannopoulos, Zhiping Weing, Kevin White, and Ross Hardison. 2014. "Defining Functional DNA Elements in the Human Genome." *Proceedings of the National Academy of Sciences* 111 (17):6131–6138.

Kohler, Robert. 1994. *Lords of the Fly: Drosophila Genetics and the Experimental Life*. Chicago: University of Chicago Press.

Minsky, Rosalind. 1990. "'The Trouble Is It's Ahistorical': The Problem of the Unconscious in Modern Feminist Theory." *Feminist Review* 36:4–14.

Müller-Wille, Staffan. 2002. "Review." *History and Philosophy of the Life Sciences* 24 (2):331–332.

Nye, Mary Jo. 2006. "Scientific Biography: History of Science by Another Means?" *Isis* 97 (2):322–329.

Phillips, Adam. 2000. "Winnicott's Hamlet." In *Winnicott Studies Monograph Series: Art, Creativity, Living*, edited by Lesley Caldwell. London: Karnac Books.

Richards, Evelleen, and John Schuster. 1989. "The Feminine Method as Myth and Accounting Resources: A Challenge to Gender Studies and Social Studies of Science." *Social Studies of Science* 19 (4):697–720.

Rose, Hilary, and Steven Rose. 2012. *Genes, Cells and Brains: The Promethean Promises of the New Biology*. London: Verso.

Rose, Steven. 1997. *Lifelines: Biology, Freedom, Determinism*. London: Penguin Books.

Rosenberg, Alex. 2006. *Darwinian Reductionism: Or, How to Stop Worrying and Love Molecular Biology.* Chicago: University of Chicago Press.

Russell, Nicholas. 1988. "Towards History of Biology in Twentieth-Century: Directed Autobiographies as Historical Sources." *The British Journal of the History of Science* 21 (1):77–89.

Schrödinger, Erwin. 1944. *What Is Life? The Physical Aspects of the Living Cell.* Cambridge: Cambridge University Press.

Selya, Rena. 2003. "Essay Review: Defined by DNA: The Intertwined Lives of James Watson and Rosalind Franklin." *Journal of the History of Biology* 36:591–597.

Shah, Esha. 2013. "Rosalind Franklin and Her Science-in-the-Making: A Situated, Sexual and Existential Portrait." *Yearbook of Women's History/Jaarboek voor vrouwengeschiedenis* 33:127–146.

Söderqvist, Thomas. 1996. "Existential Projects and Existential Choice in Science: Science Biography as an Edifying Genre." In *Telling Lives in Science: Essays in Scientific Biography*, edited by Michael Shortland and Richard Yeo, 45–84. New York: Cambridge University Press.

Söderqvist, Thomas, ed. 2007. *The History and Poetics of Scientific Biography.* Burlington, VT: Ashgate.

Söderqvist, Thomas. 2011. "The Seven Sisters: Subgenres of Bioi of Contemporary Life Scientists." *Journal of the History of Biology* 44:633–650.

Stolorow, Robert. 2011. *World, Affectivity, Trauma: Heidegger and Post-Cartesian Psychoanalysis.* New York: Routledge.

Stotz, Karola, and Paul Griffiths. 2004. "Genes: Philosophical Analysis Put to the Test." *History and Philosophy of the Life Sciences* 26:5–28.

Weinstock, George. 2007. "ENCODE: More Genomic Empowerment." *Genome Research* 17:667–668.

White, Paul. 2009. "Introduction, Special Issue on Focus: The Emotional Economy of Science." *Isis* 100 (4):792–797.

Winnicott, Donald. 1958. "The Capacity to Be Alone." *The International Journal of Psychoanalysis* 39:416–420.

Winnicott, Donald. 1965. "The Maturational Process and the Facilitating Environment." London: Hogarth Press.

Wright, Ken. 2000. "To Make Experience Sing." In *Winnicott Studies Monograph Series: Art, Creativity, Living*, edited by Lesley Caldwell. London: Karnac Books.

6

ROSALIND FRANKLIN AND HER
SCIENCE-IN-THE-MAKING

A situated, sexual, and existential portrait[1]

The story of the discovery of the double helix has become an iconic representation of the biased accreditation in science. Rosalind Franklin and James Watson emerge as pivotal anchors of this story; their stories are presented as intertwined inseparably yet antithetical to each other. The discovery of the structure of DNA was ascribed to James Watson and Francis Crick following their famous paper "Molecular Structure of Nucleic Acid", published in *Nature* in 1953. Along with Maurice Wilkins, the duo even collected a Nobel Prize in 1962 for the discovery. The controversy on Rosalind Franklin's role in the discovery erupted after the publication in 1968 of James Watson's autobiography, in which he made disrespectful remarks about Franklin as a woman and as a scientist.[2] Two biographies of Franklin have been published since Watson's autobiography, one by Franklin's close friend Anne Sayre and a second by Brenda Maddox, both of whom make a passionate effort to rescue Franklin from Watson's damaging portrayal (Sayre 1975; Maddox 2002). Maddox's biography succeeds relatively well in creating a balanced image of Franklin; however, in Anne Sayre's biography, and even more so in the debates on gender and the accreditation of science,[3] Franklin is immortalized either as an unsung hero or as a victim of discriminatory practices.[4] In these accounts, Franklin and Watson occupy opposite strands of the representation – Watson is depicted as a sexist, open-mouthed, pulse-racing, girl-mad, deceitful, juvenile boy playing with a mechanistic, masculine, "tinker-toy" model of the double helix, and Franklin is portrayed as a strong and dynamic woman and a dexterous, intelligent, overly methodical, and ethical scientist with an international reputation in crystallography who was mistreated in the men's world of science.[5] Apart from these obverse portrayals of Watson and Franklin, three questions related to Franklin's role in the discovery and her method have been debated in the historiography of the double helix: 1) Would she have ever arrived at the double helix structure by religiously following the empirical method she chose? 2) Would she have completed the final two steps needed towards interpreting her famous photograph 51 and made the discovery of the double helix herself? 3) Was she or was she not anti-helical?[6] Answers to these questions do not only crucially determine the importance of her role in the discovery but also have wider significance for the debates on gender and science, feminist epistemologies, and the relative merits of empirical/deductive versus intuitive/inductive – in other words, Franklin's versus Watson's – methods.

My contention is that these "oppositional" representations of the epistemic entities – agent, structure and method (Franklin versus Watson, helical versus anti-helical, and empirical versus intuitive) – refer to science as a product. Only in representations of the final outcome of science does it become possible for obverse images of the epistemic agents to emerge. Considering science as a product makes invisible the underlying commonality among the apparently contrasting agencies. At the same time, it underplays the complexity of the always in-the-making configurations and manifestations of the epistemic entities.

Here, I aim to engage with science as a process. I posit that the course of events leading to the discovery was a process of crystallizing "objectivity" – which means making "the cognitive and perceptual verifiable, epistemologically warranted and communicable" (Hacking 2012, 20). As discussed at length in the previous chapter, Ian Hacking explains how Kant described objective and subjective: "practical principles . . . are subjective . . . when the condition is regarded by the subject as valid only for his own will. They are objective . . . when they are recognized as . . . valid for the will of every rational being". Hacking thus defines "[o]bjective as inter-subjective" (Hacking 2012, 20). Following Hacking's definition of objectivity, I would argue that the science leading up to the discovery was a process aimed at finding the structure of DNA that was valid for all agents involved. The process of crystallizing objectivity was thus intensely intersubjective. This intersubjective drama involved intense power struggles and an intense entanglement of the epistemic agents. I aim to follow the agency of Rosalind Franklin as it emerged or unfolded during the process of intra-action or entanglement with other epistemic agents, apparatuses, concepts, and objects of knowledge during the making of her science. Here I follow the work of science historian Karen Barad, who defines intra-action as the mutual constitution of entangled agencies. While the conventional connotation of "interaction" assumes the existence of separate individual agencies that precede their interaction, the notion of intra-action means that distinct entities do not precede but emerge through the intra-action (Barad 2006, 33).

114

The purpose of this chapter is threefold. First, in focusing on the temporal unfolding of Rosalind Franklin's agency in the making of her science, I aim to make visible the underlying trope of the "scientist-self" in the portrayals of Franklin. My claim is that implied in the obverse caricatures of Watson and Franklin in the debates on gendered accreditation in science is the portrait of the scientist-self. In this projection, the scientist is an autonomous, self-determining, and self-regulated individual who is continually making rational cognitive choices. Almost all discussions on the controversies refer to both Franklin and Watson as autonomous, dispassionate, disembodied, self-willed individuals engaged in the conscious production of knowledge. By reading the history of the discovery of the double helix along with Franklin's biographies, I wish to challenge the notion of the scientist-self as an autonomous cognitive agent. In revising the notion of the scientist-self as a fundamentally feeling, suffering, and experiencing subject that emerges in the pursuit of objectivity, I aim to trace the agency of the scientist Rosalind Franklin as it unfolds on the interface of her "situatedness" (i.e., her Jewishness, her aristocratic class, her acquired French sophistication, her gender, her sexuality, and her existential self).[7] Second, I will locate the unfolding of the deeper existential and situated self of Franklin in intra-action with not only other epistemic agents but also with epistemic apparatuses and concepts. In doing so, I aim to pay attention to the material aspects of science as a process. Finally, in the conclusion, I will revisit the debates on gender and science and feminist epistemologies in the light of the revised notion of the scientist-self in entanglement with the materiality of the making of science.

A word on method is pertinent here. Biography as a genre in the history of science is not new this is also partly discussed in the previous chapters.[8] In fact, through presenting the micro-life-history of an individual, biographies shed light on the scientific collective in ways that other forms of the history of science fail to accomplish.[9] However, biographies as a source of history-writing should be regarded with some caution. Scientific biographers busy creating a "scientific Cinderella" – a hero rising from marginality to stardom with dazzling brilliance – hardly pay any attention to sexuality or to the deeper existential self.[10] Strictly speaking, Franklin's biographer Brenda Maddox does not fit into this category of biographers in the sense that she is not after creating a scientific Cinderella; however, the attention to Franklin's deeper existential, sexual, and emotional self is not the main purpose of her biography. In attempting to construct a picture of these aspects of Franklin's self, I thus run the risk of reading in-between the lines – the kind of mind reading that historians of science are wary of. In my own hermeneutic method, I have based my interpretations on the representation of the self that Franklin's close friends, family, and colleagues have described. In fact, here I do not attempt to conjure up the deeper psychoanalytical and existential subject but to construct the role of Franklin's existential self-assertion in the making of her science. My aim is to show that the scientist is not quintessentially a rational agent in pursuit of truth but to demonstrate that she tenaciously follows the path that corresponds with her self-assertion. This I aim to accomplish based on the vocabulary that Franklin herself used in the letters, the drafts and the sketches, and on the personal secrets and the metaphysical world view presented in her biographies.[11] The paper, however, remains a hermeneutic interpretation of the scientist-self in intra-action with other agents, apparatuses, and concepts in the making of her science.

"He's so middle class!"

Franklin's 27 unhappy months at King's College London are generally attributed to the fact that she was a woman scientist marginalized and mistreated in a man's world. It is frequently narrated in studies on gender and science that Franklin was wronged in this "man's world" because she was hired as an "assistant" merely to produce X-ray films supposed to be interpreted by her "boss" Maurice Wilkins. Franklin was, however, already an authority on the crystallography of coal when she arrived in London in January 1951 as a Turner Newall Fellow. This was a time when women were not employed very easily in public institutions. At King's, however, 8 out of 31 of the biophysics staff were women. Skewed as it was, the number was still exceptionally high at the time. It is also often pointed out that women were not allowed in the senior common room at King's.[12] Even the freethinking University College London had one common room only for men and another joint common room for men and women, which was known as "the joint". King's also had two dining rooms, one only for men and another for both men and women. Interestingly, though, many male scientists preferred to eat in the common dining room overlooking the Thames and refused to go to the men-only common room because it was considered a domain of "hooded crows" – Maddox does not engage with this term except by supplying the explanatory term "clerics" between brackets, possibly referring to clerics employed to train students for Anglican priesthood (Maddox 2002, 128). In all likelihood, King's was dominated by the clerics at that time; the scientists, both men and women, were accordingly relegated to a less important position. In highlighting the gender-related discrimination in narrating Franklin's story, all other forms of discriminations are constantly ignored or downplayed. However, as the "senior" scientist-men preferred to eat in the common dining room and avoided the men-only common room, which was considered a domain of "hooded crows", it could be interpreted that the discrimination was actually practiced not against women but was a matter of science versus religion.

In Franklin's legendary falling out with Wilkins, gender had played a similarly ambivalent role. Rosalind herself came from a long line of illustrious Anglo-Jewish scholars and leaders. Her family owned a private merchant bank as well as a publishing company and had been wealthy and influential in England for almost two centuries. Returning from France – where she worked between 1947 and 1950 and where she was surrounded by intellectual Jews – to King's in 1951, Franklin was very conscious of her Jewishness, but it was not something that she openly identified with. In England while working on DNA her upper-class identity in combination with her sexuality and her existential self-assertion unfolded in a way that crucially determined the course of the events, her connection with Wilkins, and her role in the discovery.

During these 27 months, Franklin frequently complained about the mediocrity around her. Her complaints were informed by her upper-class background, which distinguished her from the rest of the staff. One of the colleagues at King's described her in the following words,

> I always supposed her family was rich, though she never talked about it – she really stood out very much here where most of the other with very few exceptions came from ordinary background middle-class or in some cases, I suppose, lower than that. Really, the word is aristocratic – she looked like an aristocrat and she acted like one . . . just the way she spoke, there were people at that time who sneered at the upper-class way of speaking, and really *hated* it (italics in the original).
>
> (Maddox 2002, 127)

Franklin's accent was the fault line of the class divide and worked both ways. Populated by people with less than "Received Pronunciation", King's was no Oxbridge. For the upper-class Anglo-Jewish woman with acquired French tastes and sophistication, this meant that she was surrounded by people who were, in Franklin's own words, "intellectually second rate" (Maddox 2002, 127). The director of the laboratory John Turton Randall was actually a son of a market gardener who made his way up through grammar school to the University of Manchester. Even after acquiring an influential position at King's, he always remained conscious of the "rough edges" of his northern ways and experienced difficulties in becoming fully accepted in the "smooth south" (Maddox 2002, 131). But Randall was not the one with whom Franklin ran into serious trouble. It was the connection with Maurice Wilkins that became a source of multiple tremors which not only shook their respective work-lives but, travelling as far as Cavendish to Watson and Crick, crucially shaped the entire process of science leading up to the discovery of DNA's double helix.

Wilkins was born in New Zealand to Anglo-Irish parents and grew up in a liberal Unitarian tradition with family members closely involved in promoting education among women; his father was founder of Working Men's college. Both Wilkins and Franklin intensely disliked the idea of science for profit. Politically, and even in terms of their social backgrounds, Franklin and Wilkins were in tune, and Maddox feels that "the ingredients for many a laboratory romance were there" (Maddox 2002, 145). According to Maddox, Wilkins was almost half in love with Franklin, except that temperamentally they were opposite. But more importantly, in my reading, it was Franklin's upper-class consciousness and her sexuality that "rejected" Wilkins as a prospective suitor and as a colleague with whom to collaborate.

Rather than Wilkins, the brilliant (and seductive) Monsieur Mering, who was Franklin's boss at the French government laboratory where she worked as a researcher between 1947 and 1950 before joining King's, was her ideal. Mering was a Russian-born Jew who remained in hiding during the years when the Nazis occupied Paris and returned to Paris as a refugee after the war. Maddox describes Mering as a "seductive charmer who plays on any woman's susceptibility". All the girls in the lab were in love with him and so was Franklin (Maddox 2002, 96). Apparently, Mering was also in love with Franklin, but their relationship remained platonic owing partly to Franklin's puritanical upbringing which meant that she would have not accepted a married man with other known mistresses as a lover, and partly to what Maddox repeatedly, but vaguely, designates as her emotional naivety and sexual inexperience. Adrienne Weill, Franklin's close French friend, described Franklin's feeling for Mering as "deep, genuine and somewhat tragic" (Maddox 2002, 168) because Franklin retained these feelings until the last day of her life without reciprocation. Her stay at King's was also the period in which she struggled to come to terms with her feelings for Mering, who refused to engage with her after she left his laboratory to join King's. Despite Mering's indifference, Franklin continued to acknowledge him, and only him, in all her published papers on coal.[13]

The fact that Franklin remained under the charismatic spell of Mering for most of her life had another side: he was the standard against which she measured every man in her life. She could only respect men who were charismatic, strong, decisive, protective, and had something to teach her, like Mering (and later J. D. Bernal) (Maddox 2002, 146). She described her colleagues at King's in rather harsh words in a letter to her close friend Anne Sayre in March 1952:

> The very young are thoroughly nice but none of them are brilliant. . . . And the other middle and senior people are positively repulsive. . . . The other serious trouble is that there isn't a first class or even good brain among them – in fact nobody with whom I particularly want to discuss anything, scientific or otherwise, and I so much prefer to work under somebody who commands my respect and can offer some encouragement.
>
> (Maddox 2002, 172)

Only Professor J. D. Bernal commanded her respect, and, responding to her friend Anne Sayre's reserve for Bernal's politics, she declared that "whatever one may have against the man, he's brilliant" (Maddox 2002, 173).

In comparison, Wilkins was not charismatic, strong, and decisive but slow and defensive, and in the face of conflict he became expressionless and quiet. Besides this, he did not even know the basics of chemistry, which increased her scorn for him. Franklin summed up her disapproval of Wilkins to her close friend as "He's so middle class!" (Maddox 2002, 147). This shows that Franklin's upper-class consciousness, her sexuality, and her emotional need to associate with strong, protective men shaped her science more than her cognitive scientist-self.

Wilkins was a Cambridge graduate who spent much of the war in Berkeley, California, working on the Manhattan project. By the time Franklin arrived at King's, Wilkins had been working on nucleic acid for several years, along with a PhD student, Raymond Gosling (who eventually worked closely with Franklin and whom she supervised). Together, Wilkins and Gosling had made some first attempts at taking X-ray pictures of DNA fibres, which were described as remarkably unique. But Wilkins and Gosling were not accomplished crystallographers as Franklin was. Nevertheless, they improvised an impressive apparatus by using a bent paperclip as a frame to mount the nucleic acid fibre on, passing saturated hydrogen through a tube to keep the fibre moist and sealing the tube with a condom to achieve vacuum. This was the first set of pictures that showed the famous x-shape which was the clear indication that DNA had a crystalline structure. On seeing the pictures and the distance between the spots on the X-ray film, a colleague, Alec Stokes, speculated that the blank spaces along the length of X meant a helix shape. The Wilkins, Gosling, and Stokes trio had thus already floated the epistemic concept of the structure of DNA as a helix by the time Franklin arrived at King's. These pictures owed their clarity to the sample of DNA prepared and distributed by Professor Signer; Wilkins had also managed to acquire some. The Signer DNA had special properties – touched by a glass rod, it could be pulled out into long fibres or strings, and it had higher molecular weight. These special properties were very important for achieving the crystalline character necessary for X-ray filming. Another specimen of DNA prepared by Erwin Chargaff did not crystallize this clearly. Wilkins had also figured out that when Signer DNA was moistened it could be extended to double the original length – a process they called "necking". Both Wilkins' and Gosling's pictures of the X-shape and their experiments with necking became precursors to the famous wet and paracrystalline B form and the dry and crystalline A form of DNA that Franklin and Wilkins later experimented with in their own separate ways. In addition to the high quality of the DNA, Wilkins had also managed to acquire the prototype of a Werner Ehrenberg fine-focus X-ray tube which made it possible to photograph a single DNA fibre of one-tenth of a millimetre and to control the humidity inside the tube (Olby 1994, 331–336).

Wilkins first presented his "helices" to an international audience in a conference on large molecules held at the Marine Biological Station in Naples (Maddox 2002, 141). In the audience was Dr James Watson. Wilkins' helix gave him the confidence that DNA molecules had a symmetrical structure which could be studied. Besides Wilkins, other sources were presenting DNA as a helix. Everyone at King's and Cavendish had read Sven Furberg's thesis, a young Norwegian researcher whose model of an arc of a helix of DNA showed the correct place of sugars and bases (Olby 1994, 336–338). Moreover, in the spring of 1951, Linus Pauling, an authority on proteins, made the discovery that the regular structure found in proteins had a shape of an alpha helix (Olby 1994, 278–289). Around the time Franklin arrived at King's and began her work, "helices were in the air" as Francis Crick later described it.

I narrate this story told and retold many times by excellent historians to make the point that by the time Franklin arrived in the crystallography of DNA, helices had acquired broad approval in the collective thought; they had not yet been elevated to the level of objective sanctity, yet, to take a non-helical stance would have needed a lot of courage and, in a certain sense, would have meant a heresy. Even long after the discovery of the double helix, the helical prophecy held sway to the extent that it became important for Franklin's best friend and biographer, Anne Sayre, and her close friend and colleague, Aaron Klug, to rescue Franklin from being anti-helical. Sayre refers to Franklin's notes, cracked by Klug, to show that she was only temporarily anti-helical and that she was actually only two steps away from the thesis of the double helix (Sayre 1975, 120–136; Klug 2004).

Sayre's and Klug's Franklin keeps the "myth of personal coherence" intact. In both these accounts, Franklin is a quintessential scientist-self making deliberate cognitive choices by carefully weighing the evidence. This projection of the scientist-self motivated only by consciously chosen cognitive pursuits hides the underlying emotional and existential drives of self-assertion. By engaging with Franklin's prank in July 1952 in which she sent around invitations for the funeral of the helix, in the following section, I wish to hazard a few remarks about the way in which Franklin's helical and anti-helical oscillations, an integral part of her science, were expressions of her "struggle for self-assertion" and her quest for "existential authenticity" (Söderqvist 1996, 27, 66).

"The DNA helix died!"

On joining King's, Franklin's first job was to redesign and make the photography apparatus work in a way that a single DNA fibre could be mounted for long hours of X-ray exposure. In the first six months after her arrival she expertly handled controlling the humidity inside the tube by using salt solutions, achieved vacuum by a specially designed pump, and designed and mounted a specific tilting micro-camera. Her expert ways impressed almost everyone, including Wilkins. She had not yet started the actual filming when a row with Wilkins erupted, prompted by Wilkins' presentation on his helical thesis at the conference organized at Cavendish in June 1951 (Maddox 2002, 148–149). Wilkins showed his X-ray pictures, taken together with Gosling, to argue yet again that the X-shape implied a regular crystalline structure and strongly suggested a helical shape. Wilkins' presentation irked Franklin, for here he was intruding in her territory – the X-ray work of DNA structure. She stopped Wilkins outside the presentation venue and demanded in a "firm and deliberate tone" that he give up X-ray work and "go back to the microscopes" (Maddox 2002, 149). Much has been said about why this happened and Maddox, based on Wilkins' own account, attributes this to laboratory politics – i.e., Randall wanting to remove Wilkins from the high-profile DNA work and to insert himself through working with Franklin. He was the one who made Franklin believe that the DNA work belonged to her and Gosling alone. Wilkins was shaken by her reaction but kept attempting to bridge the distance between them, both personally (even bringing chocolates for her in one instance) and academically. In the meantime, Franklin, with Gosling, succeeded in identifying the separate A form (crystalline and dry) and B form (paracrystalline and wet), showing distinct X-ray images. The pictures of the B form, in particular, gave the first clear evidence of how the molecule opens up to replicate itself. On seeing the pictures of the B form, Wilkins made yet another attempt to show Franklin that her pictures matched beautifully with the helix equations worked out by Stokes (Stokes humorously called them "waves on Bessel-on-sea" – a play on Bessel functions being mathematical formulas used for structural calculations, and the beautiful Sussex town of Bexhill-on-Sea) (Maddox 2002, 152). This was another invitation for collaboration which Franklin turned down; neither did she find Stokes' wave patterns "beautiful". In October, the boundaries were finally marked clearly through the intervention of the laboratory director, John Turton Randall: using Signer's DNA, Franklin would continue to focus on the A (dry) form and Wilkins using Chargaff's DNA would focus on the hydrated B form (Maddox 2002, 155). The new, expertly constructed X-ray equipment was designated for the exclusive use of Franklin. From this moment on, Wilkins was out of the "DNA race" because he was neither an expert crystallographer nor did he have access to high molecular weight Signer DNA or to X-ray equipment. Thus, due to the fact that Franklin had declined to collaborate with him, Wilkins was marginalized from the DNA work which he had founded at King's. By this, I do not wish to imply that Franklin purposely worked against Wilkins; rather, my aim is to show the complexity of the entanglement of the epistemic agents and the influence of the unintended consequences on the final outcome, which could only be made visible by exploring science-as-a-process.

Franklin chose the crystallographer's method of Patterson function analysis to come up with the structure of DNA. A Patterson map was like a contour map showing the peaks and troughs of the X-ray diffraction image. This was a laborious method, requiring methodical registration of intensities of spots on the X-ray picture, which were then transferred onto a map through a series of mathematical calculations. It was acknowledged that the Patterson method needed not only patience and intense concentration during long hours of hard work but also a bit of luck. This exercise proved far more difficult than expected. Not only had Franklin never done this before, but no one had ever tried cylindrical sections before. In Gosling's words, "it was trying to do a three-dimensional jigsaw puzzle" (Maddox 2002, 170). A colleague described Franklin and Gosling working on the Patterson analysis "week after week, buried in paper, getting nowhere, buried in tables, slide rules and desk calculations requiring extreme concentration". "We spent ages, we had to think in three dimensions", said Gosling (Maddox 2002, 183–184). Choosing the Patterson method meant, on the one hand, that she was religiously following the methods in crystallography – creating Patterson maps was what "crystallographers would do" (Maddox 2002, 173). On the other hand, this meant that she was going against the current, and in doing so, she was indeed, if not rejecting, then at least questioning the helical hypothesis. The fact that Franklin did not succeed in demonstrating a viable non-helical alternative structure of DNA could be interpreted, as Watson did, as showing that the impossibility of the "choice" was written all over it, and in choosing so, Franklin was an "unimaginative" and "uncreative" scientist. Conversely, however, Anne Sayre strives to prove that Franklin's choice actually signified "objective" and "empirically sound" and "methodical" science and, in choosing so, Franklin was "one of the world's great experimental scientists" (Sayre 1975, 146).

In my interpretation, Franklin's choice of the Patterson method for her DNA work was a classic Lacanian paradox of the subject – how every turn of logic reveals its illogicality (Žižek 2006). What looks like a "free choice" of method on Franklin's side was actually a moment of no choice. For the scientist, no fear is stronger than the "anxiety of influence", the horror of finding one's work only a copy, the fear that one's work will be forgotten or ignored or that nothing distinctive will be ever found in it (Bloom 1973, 80). Franklin was not the kind of person who plucked the low-hanging fruit of helices floating in the air. She was a determined scientist whose science would always have to be distinctly original. She was willing to go a long way and, retrospectively speaking, she even paid with her own life in achieving this goal of originality and precision. A colleague at King's once found her in the basement X-ray room working late at night struggling to fix the tilting camera, which could be done only when the X-ray beam was on, and she was standing in the beam without the protective lead aprons, too eager to fix the camera (Maddox 2002, 144). Later, she worked on the structure of TMV and polio, and we know that Franklin died at the young age of 37 due to ovarian cancer and polio. What looks like a free choice of rejecting the collaboration with Wilkins and, along with him, the helices hanging in the air was the moment of choosing her own authentic existential self.

The notion of existential choice is defined by Kierkegaard in his Either/ Or, in his discussion of the choice between ethical life, in which questions of good and evil guide the choices in life, and aesthetical life, in which the search for beauty, truth, joy, and symmetry becomes subordinated to ethical priorities. For Kierkegaard, the ethical individual not only knows himself but actually "chooses his self" in the sense that the individual is determined by his own necessity.[14] The question of choice and freedom is crucial here. To explain this further in terms of Lacan interpreted by Žižek, the subject only retroactively posits the causes of its own existence; freedom is also posited only retroactively. What's more, the subject does not select from an infinite list of possibilities (as if choosing a product in the supermarket) but chooses the "necessities that will determine itself". For Žižek, this is the Lacanian paradox of the subject, a parallax moment, the choice that is no choice, or the choice that is nothing but choosing the necessity of one's own existence and hence no choice (Žižek 2006, 25–26).

I read Franklin's crusade to question helices, her refusal to collaborate with Wilkins, and her brave embracing of the impossibility of the Patterson method as choosing the necessity of her scientist-self. To interpret her choices in the narrow sense of whether she was helical or anti-helical would be to typify her purely in terms of rational and cognitive agency – resulting in a portrayal of Franklin either as "stupid" (Watson) or "brilliant" (Sayre). In either case she would turn out to be a(n) ir/rational cognitive agent mechanically weighing the evidence and emerging as a good or bad scientist based only on the outcome of her science. Exploring science as a process has revealed, however, the compelling influence of the deeper existential drives which condition the subject to act in such a way that questions of right and wrong are not consulted, calculated, or pondered upon.

This deeper, compelling existential drive was also manifested in a prank that Franklin masterminded. On 18 July 1952, "with the air of freedom in her nostrils" (Maddox 2002, 184), as her biographer Maddox presents it, Franklin wrote a funeral card for the DNA helix. Signed R. E. Franklin on one side and R. Gosling on the other, the notice was handwritten by Franklin in black capital letters, which read,

> It is with great regret that we have to announce the death, on Friday 18 July 1952 of D. N. A. Helix (crystalline). Death followed a protracted illness which an intensive course of Besselises [sic] injections had failed to relieve. A memorial service will be held next Monday or Tuesday. It is hoped that Dr. M. H. F. Wilkins will speak in memory of the late Helix.
>
> (Maddox 2002, 135)

Maddox tries to convince the readers that this was just a joke, and points out that there are debates about the extent to which this funeral invitation was distributed (Elkin 2003, 46). Historian Robert Olby interprets this funeral note as a sign that Franklin was anti-helical (Olby 1994). In my reading, following a Freudian-Lacanian frame, this prank was saturated with the investment of Franklin's subconscious desire. In other words, the demise of the helix was a fantasy from which she derived enjoyment, which, in turn, provided "a necessary vehicle to organise her reality into a coherent whole" (Žižek 1997). In Lacan's work, fantasy narratives structure real practices by providing a convincing explanation of the lack of "total enjoyment" (*jouissance*).[15] This lack is articulated in terms of an absence of some form of utopia. In her prank, Franklin constructs a world without helices. And this world of Franklin's was not a world that emerged logically from the empirical facts – as both Watson and Sayre in contrasting ways would have it. It was actually the other way round. This prank shows that her subconscious world of desire and fantasy was structuring her quest for "facts". As Michael Polanyi puts it, rational knowing involves the existential participation of the knower – that is to say, scientific knowledge involves existential choices, which for Polanyi means that existence precedes essence (Polanyi 1958). This prank shows that Franklin's conceptual world answered to her emotional longing and that her theoretical world was isomorphic with the world emotionally sought.[16]

What is partly inexplicable is why Franklin's emotional world sought for a world without helices. I surmise that this was driven partly by her search for existential authenticity and the quest for originality and uniqueness in her scientific work, and partly by the connection of the helix with Wilkins and Franklin's own conflicting association with Wilkins – this unfortunately could only be explained in terms of transference, which is beyond the scope of this chapter.

"It's so beautiful, you see, so beautiful"

The last episode of the discovery of DNA's structure was, as Wilkins aptly described, "a helical rat race" (Maddox 2002, 208). I will be brief about the last chapter of the DNA saga because excellent accounts have been written of the hour-to-hour development of this story of deceit and behind-the-scenes manipulation.

Watson and Crick decided to build a model one week after the colloquium on nucleic acid at King's in November 1951 at which Wilkins, Stokes, and Franklin spoke about their work on the structure of DNA. Watson had taken no notes at the colloquium so, based only on the memory of what he heard on stereo-chemical details, they put together a three-chain model of a helix with the phosphates on the inside and the bases on the outside. The team at King's – Wilkins, Franklin, Gosling, and two other colleagues – were invited to see the model. Without wasting any time on greetings, Franklin immediately pointed out that the calculations were grossly incorrect and pointed out several mistakes in a matter of minutes – the phosphates had to be on the outside, water had to be shown ten times higher than was represented, sodium ions had to be on the inside and, if they were encased in water as Watson and Crick had suggested, they would be unavailable for binding, so that was wrong, too. Watson described her as "pugnaciously assertive". Franklin was a scientist of precision – she would never speak until she was absolutely certain – and she was positively snorting about the obvious mistakes the duo had made. The model made it more than clear that neither Watson nor Crick knew their basic chemistry. And this was Franklin's powerful response to the patronizing attitude that Watson and Crick had adopted with her (Maddox 2002, 179).

A year or so later, the situation markedly changed when Watson and Crick built their second model. As is well-known, Watson's deceitful peek at Franklin's famous photograph 51 played a decisive role in this last chapter of the rat race. Photograph 51 was, curiously enough, an unintended consequence of Franklin and Gosling's laborious X-ray work. The X-ray filming required as much as 100 hours of exposure. During one such exposure the single DNA fibre mounted for filming changed its form from crystalline A to paracrystalline B. Photo 51 was an accidental, but the clearest, best-ever exposure of the B form, which showed a clear X – unquestionably a helix. Franklin kept this photo in her drawer and continued working on the A form to which she had been assigned – as agreed, Wilkins was handling the B form. It was Gosling who brought that picture to Wilkins in January 1953, eight months after Franklin and Gosling's combined work unintentionally produced this picture; Gosling still retained his loyalty to his former colleague and advisor. Wilkins showed this picture to Watson, without Franklin's (or even Gosling's) knowledge, causing Watson's "jaw to drop, mouth to fall open, and pulse to rise" (Watson 1968, 24). As has been told many times, Watson apparently cracked the code of the double helix in the train on his way back to Cambridge.

"Rosy's Parlour"

In March 1953, Franklin left King's and joined J. D. Bernal's group at Birkbeck. In the annual Christmas celebrations of the physics and biophysics units at King's that year, "Rosy's Parlour" was one of the themes. The following is what was printed in the program's advertisement,

> Best crystalline nucleic acids:
>   Dehydrated, Uviated, de-Rosieated
> (Frustrated Exports), Infraredded, X-rayed
>   And other wise Maltreated
> Also: Hands Read, Bumps Told, Patterson Diagrams
>   Of your Future (and Past)

By Request
MADAME RAYMONDE FRANKLINE
Clairvoyante
(Maddox 2002, 232)

Maddox interprets this as a celebration of the fact that Franklin was gone. I read it in entirely the opposite way. Jocular commemorations have culturally been part of the Christmas celebrations of not only the biophysics group at King's but of many other places, for example, Cold Spring Harbor and Niels Bohr's institute in Copenhagen. Popular modes of celebration in the scientific culture of the time, they combined the formal scientific advances and problems in the field with more spontaneous emotive aspects, which could be interpreted "as stories they tell themselves about themselves" (Beller 1999, 252). These events, including those at King's, had a carnivalesque character that in a Bakhtinian sense simultaneously celebrated and criticized, affirmed and denied; they represented authority, hierarchy, and order while providing a subversive space to challenge these; they provided an outlet for accumulated tensions and frustrations; and they provided space to say the unsayable but at the same time they recorded the collective lived experience. The fact that Rosalind Franklin and Raymond Gosling were the subject of jocular celebration and not Wilkins and Stocks or even Watson and Crick's famed double helix showed that Franklin and Gosling's X-ray films of DNA were at the centre of the live nerve of the collective emotive experience at King's. Franklin emerges here neither a hero nor a villain nor a victim. And the mocking humour concerning her represents the tension between genuine affection and unbearable frustrations – it represents the paradoxical persona of Franklin imbued with ambiguity, authority, hierarchy, vitality, frustration, courage, and rock-solid conviction – a celebration of the decentred "human" person that she was. This jocular commemoration confirms that she put an indelible mark on the collective emotional experience at the King's – an honour which was only hers.

Conclusion

For a long time, feminist epistemologies have been engaged with theorizing the relationship between objectivity and subjectivity, in which the idea of what can be counted as objective science and who is the knowing subject have been revised from their positivist and foundationalist predecessors.[18] The idea of the paradigmatic knower as an unbiased, disembodied individual purified of all passions and emotions has been centrally questioned in feminist epistemologies. Also, the idea of science as objective and value-free knowledge produced by a rational and autonomous knower has been widely challenged. But what the theory of the subject is that should replace this empiricist and positivist science has long been debated in feminist epistemologies. For the sake of brevity, later I briefly discuss three such approaches that link objectivity with subjectivity. My broad classification is meant to capture the spirit of the differences in the approaches as they evolved in the feminist epistemologies; it is by no means exhaustive. The first approach of standpoint theory revolutionized the question of method in feminism, constructing the "ideal" epistemic agent not as the unconditioned and purified subject but as a subject conditioned by social experiences of domination. Some feminists thus valorize the knowledge and perspectives available to those who occupy a disadvantageous position in the multidimensional power grid. However, the position of marginality in standpoint theory does not automatically earn a privileged epistemic position; in order to become a standpoint of epistemic advantage, it has to be organized, through deliberate political and theoretical struggles, into a conscious political project. For such a political project of a standpoint to emerge, subjective experiences and knowledge, both affective and cognitive, from the position of marginality are nevertheless privileged (Harding 1986). There are two problems with standpoint epistemologies. First, they assume that the subject position of marginality is *a priori* available and that it remains so, unchanging for the political project of the privileged epistemic position to form around it. Second, the epistemic method is assumed to be a product emerging entirely out of the complex and socially and politically manifested power dynamics among the subjective positions.

The second approach in feminist epistemologies linking objectivity with subjectivity can be located in the notion of situation. The bias emerging from the situated location of the knowing subject is widely recognized. The subject is formed in a situated location by a range of power asymmetries, i.e., gender in combination with other structural categories, such as race, class, and ethnicity (Longino 1993, 109). It is at the specific situated location – at the locus of experience – that gender becomes an important analytical category (Nelson 1990). Knowledge is thus integral to the lives of social agents. Evelyn Fox Keller's classic work of biography on Barbara McClintock, discussed at length in Chapter 5, figures in this category (Keller 1983).

Lastly, the third approach focuses on science as a process and, in different ways, emphasizes relationality and narrativity. Helen Longino's "norms of transformative criticism" propose an understanding of the way in which "complexly conditioned subjectivities are expressed in action and belief" and the way science is constructed by relations of interdependence, by "individuals in interactions with one another modifying observations, theories and hypothesis, and patterns of reasoning" (Longino 1993, 109). The knowing subject is thus in dialogue with the epistemic community. In fact, the epistemic subjects are the communities (Longino 2001). More recently, Karen Barad talks about the mutual constitution of the entangled epistemic agencies – agents, apparatuses, concepts, and objects – the approach adopted in this chapter (Barad 2003). In this approach, the knowing subject is heterogeneous, multiple, contradictory, and incoherent, not given but emerging in the process of the making of science.

How does the unfolding agency of Rosalind Franklin in the making of the science of DNA engage with this repertoire of feminist epistemologies? I have the following observations to make. First, tracing the unfolding of the agency of Rosalind Franklin over the temporality of the entire process of the discovery of the double helix fundamentally challenges the simplistic images of Franklin as either a victim or a hero. The complex micropolitics of power that unfolded during the process make it difficult to find one structural position of marginality or domination that can be ascribed to Franklin. Her own personal association with other epistemic agents, concepts, and objects was fraught with contradictions in such a way that her agency occupied positions of domination and marginality at the same time during different temporal and epistemic locations in the process. Locating the process of the DNA discovery as integral to her life experience, including her sexual and existential self, produces a messier picture than the one produced by studies concerned with the gendered accreditation of science.

Second, the portrait of Franklin's agency unfolding in the making of her science challenges the way her scientist-self is usually characterized in the debates on gender-based discrimination in science. Her portrayal as either a hero or a victim validates the myth of personal coherence. It retains the notion of the scientist-self as an autonomous and wilful agent, making rational cognitive choices. A picture of Franklin's science-in-the-making shows that the choices she made were hardly the conscious, deliberate choices usually ascribed to a rational and calculating agent in pursuit of "truth". Her "choice" of the Patterson method and her refusal to collaborate with Wilkins and Stokes stemmed from the tenacious agency of the self, shaped on the interface of her situatedness – her gender, class, and ethnicity – and her deeper affective and existential assertions. The discourses on the success or failure of Franklin's science and, correspondingly, the description of her science as brilliant or stupid overlooks the compelling role of these deeper drives. What's more, discourses on success or failure ignore the process of the unfolding of the epistemic agency in intra-action, in entanglement, with the agencies of other agents, objects, and concepts. Franklin's conflicting relationship with Wilkins, her arduous following of the epistemic method of Patterson maps, her expertly constructed relationship with the epistemic apparatuses – the camera, the tube, the sample DNA, her unintended photograph 51, and her ambivalent relationship with the epistemic concept of the helix – all these unfolded in a complex, messy, and unintended fashion and shaped the outcome of the end-product of her science.

In conclusion, in following Franklin's science-in-the-making, I argue that scientists' particular ways of being and belonging in the world, in intra-action with other epistemic agents, objects, and concepts, pioneer the structuring of rational and cognitive thought.

Notes

1 This chapter was first published as Shah, E. 2013. "Rosalind Franklin and Her Science-in-the-Making: A Situated, Sexual, and Existential Portrait." *Yearbook of Women's History/Jaarboek voor vrouwengeschiedenis*, in Special Issue on *Gender and Genes*, 33, 127–146, and is republished here with the kind permission of the editors.

2 Watson constructed Franklin as an aggressive, unattractive, unfeminine, and humourless "Rosy" whose science was unimaginative (Watson 1968). Both Sayre's and Maddox's biographies point out that Watson and Crick constructed the model after having deceitful access to Franklin's data and her X-ray pictures. It is also important to point out that both Maurice Wilkins and Francis Crick did not agree with Watson's portrayal of Franklin and angrily objected to the publication of the autobiography by Harvard University Press. Watson had made damaging remarks not only about Franklin but many other scientists of reputation he had worked with. Harvard University Press did not publish the book, but it was later published by Athenaeum Press in New York and Weidenfeld and Nicolson in London with Franklin's original portrayal intact. The book has since been a best-seller, has run into many editions, and, since, Watson has published two more personal accounts of the discovery of the double helix, the most recent of which was *The Annotated and Illustrated Double Helix* published in 2012. For the story behind the original publication and Watson's interview on the publication of the latest version, see McKie (2012).

3 Throughout the chapter I have made a distinction between the debates on gender and accreditation in science and feminist epistemologies. The former concerns the gender-based discrimination in the accreditation of science, meaning the way in which women's work is devalued, stolen, not acknowledged, and not rewarded in the men's world of science; the latter addresses the way androcentric bias shapes scientific method.

4 For instance, reviewing Brenda Maddox's biography Sara Delmont acknowledges the role of class, race, London versus provinces, Europe versus England, and Russia and versus America in the complex story of Franklin's science-in-the-making, but in the rest of the review she focuses only on the gender issue (Delamont 2003).

Hilary Rose projects Franklin as an example of gendered science, observing that "the woman scientist is always working in the men's laboratory" (Rose 1983). For yet another picture of Franklin as a victim – her early death, her gender, the theft of her pictures and data, and the character assassination by Watson – see Elkin (2003). Rapoport does the opposite, turning the victim into a hero (Rapoport 2002). For a portrait in which Franklin emerges, yet again, as a victim of sexism, anti-Semitism, and Christianity, see Rose (1994). Franklin's case is cited many times as an example of gender-bias in science; see Noble (1992) and Keller (1985).

5   For a detailed discussion on the construction of these contrasting images in Sayre's and Watson's biographies and in studies on gender and science, see Richards and Schuster (1989).

6   Franklin's close friend and colleague Aaron Klug and Anne Sayre have passionately argued that Franklin was only temporarily anti-helical. Science historian Robert Olby has given a detailed account of how and when Franklin changed helical position vis-à-vis A and B form of DNA. See Olby (1994, 323–423); also see Klug (2004) and Sayre (1975, 120–136).

7   Ideally, I would have liked to engage similarly with the unfolding of the agencies of three other epistemic agents – Maurice Wilkins, James Watson, and Francis Crick – and their mutual entanglement, but this would warrant a separate chapter.

8   For a spirited defence of biographies as a mode of scientific history, see Nye (2006). See also Söderqvist (1996, 2007, 2011).

9   For example, in the fascinating biography of Darwin, the biographers show how his hatred of slavery shaped Darwin's views on human evolution (Desmond and Moore 1991). Another example is the way Katherine Hayles uses Norbert Wiener's autobiography to argue that the scientist's tormented hallucinations were integral to his science of cybernetics, thereby challenging the idea of the scientist-self as primarily a rational and cognitive agent (Hayles 1999, 84–112). Further, in one of the rare studies that focuses on sexuality and science, Elizabeth Wilson uses biographical literature to show the way affect, sexuality, and brilliant scientific work overlapped in the early history of Artificial Intelligence (Wilson 2009).

10 For further discussion on this point, see Abir-Am (1991).

11 For the further discussion on the role of existential choices in history of science, see Söderqvist (1996).

12 Almost every representation of Franklin as a victim of a man's world has mentioned this point. See the reference note 3.

13 Maddox clarifies twice in her biography that Franklin's family did not agree with Anne Sayre's view that Franklin had deep feelings for Mering. Sayre and Maddox, however, clearly mention and discuss this.

14 As discussed in Söderqvist (1996, 72–74).

15 In Lacanian psychoanalysis, *jouissance* is a pre-symbolic real enjoyment which is always posited as something lost, as a lost fullness – the part of ourselves that is sacrificed when we enter the symbolic system of language and social relations. It refers to the desire of fullness that is ultimately impossible. For further discussion, see Stavrakakis (1999, 41–42).

16 For further discussion on how the conceptual world is isomorphic with the world emotionally sought, see Söderqvist (1996, 69).

17 I here explicitly discuss feminist epistemologies and hence some of the promi-
nent methodological trends in STS, for instance, constructivism and actor-
network theories are not discussed.

References

Abir-Am, Pnina. 1991. "Nobelesse Oblige: Lives of Molecular Biologists." *Isis* 82 (2):
326–343.

Barad, Karen. 2003. "Posthumanist Performativity: Toward an Understanding of How
Matter Comes to Matter." *Signs: Journal of Women in Culture and Society* 28 (3):
801–831.

Barad, Karen. 2006. *Meeting the Universe Halfway: Quantum Physics and the Entanglement of Matter and Meaning.* Durham, NC: Duke University Press.

Beller, Mara. 1999. "Jocular Commemorations: The Copenhagen Spirit." *Osiris*
14:252–273.

Bloom, Harold. 1973. *The Anxiety of Influence.* New York: Oxford University Press.

Delamont, Sara. 2003. "Rosalind Franklin and Lucky Jim: Misogyny in the Two
Cultures." *Social Studies of Science* 33 (2):315–322.

Desmond, Adrian, and James Moore. 1991. *The Life of a Tormented Evolutionist:
Darwin.* New York: A Time Warner Company.

Elkin, Lynne. 2003. "Rosalind Franklin and the Double Helix." *Physics Today* 56 (3):
42–48.

Hacking, Ian. 2012. "Objectivity in Historical Perspective." *Metascience* 11:11–39.

Harding, Sandra. 1986. *The Science Question in Feminism.* New York: Cornell University Press.

Hayles, Katherine. 1999. *How We Became Posthuman.* Chicago: University of Chicago Press.

Keller, Evelyn Fox. 1983. *A Feeling for the Organism: The Life and Work of Barbara
McClintock.* New York: W.H. Freeman and Company.

Keller, Evelyn Fox. 1985. *Reflections on Gender and Science.* New Haven, CT: Yale
University Press.

Klug, Aaron. 2004. "The Discovery of the DNA Double Helix." *Journal of Molecular
Biology* 335:3–26.

Longino, Helen E. 1993. "Subjects, Power, and Knowledge: Description and Prescription in Feminist Philosophies of Science." In *Feminist Epistemologies*, edited
by Linda Alcoff and Elizabeth Potter, 101–120. London: Routledge.

Longino, Helen E. 2001. "Can There Be a Feminist Science?" In *Women, Science,
and Technology: A Reader in Feminist Science Studies*, edited by Mary Wyer, Mary
Barbercheck, Donna Giesman, Hatice Ozturk and Marta Wayne. New York:
Routledge.

Maddox, Brenda. 2002. *Rosalind Franklin: The Dark Lady of DNA.* London: Harper
Collins Publishers.

McKie, Robin. 2012. "DNA Pioneer James Watson Reveals Helix Story Was
Almost Never Told." *The Guardian*, 9 December.

Nelson, Lynn. 1990. *Who Knows: From Quine to Feminist Empiricism*. Philadelphia: Temple University Press.

Noble, David. 1992. *A World without Women*. New York: Oxford University Press.

Nye, Mary Jo. 2006. "Scientific Biography: History of Science by Another Means?" *Isis* 97 (2):322–329.

Olby, Robert. 1994. *The Path to the Double Helix: The Discovery of DNA*. New York: Dover Publications, Inc.

Polanyi, Michael. 1958. *Personal Knowledge: Towards a Post-Critical Philosophy*. Chicago: University of Chicago Press.

Rapoport, Sarah. 2002. "Rosalind Franklin: Unsung Hero of the DNA Revolution." *The History Teacher* 36 (1):116–127.

Richards, Evelleen, and John Schuster. 1989. "The Feminine Method as Myth and Accounting Resource: A Challenge to Gender Studies and Social Studies of Science." *Social Studies of Science* 19 (4):697–720.

Rose, Hilary. 1983. "Hand, Brain, and Heart: A Feminist Epistemology for the Natural Science." *Signs* 9 (1):73–90.

Rose, Hilary. 1994. *Love, Power, Knowledge*. Cambridge: Polity Press.

Sayre, Anne. 1975. *Rosalind Franklin and DNA*. New York: W.W. Norton & Company.

Söderqvist, Thomas. 1996. "Existential Projects and Existential Choice in Science: Science Biography as an Edifying Genre." In *Telling Lives in Science: Essays in Scientific Biography*, edited by Michael Shortland and Richard Yeo, 45–84. New York: Cambridge University Press.

Söderqvist, Thomas, ed. 2007. *The History and Poetics of Scientific Biography*. Burlington, VT: Ashgate.

Söderqvist, Thomas. 2011. "The Seven Sisters: Subgenres of *Bioi* of Contemporary Life Scientists." *Journal of the History of Biology* 44:633–650.

Stavrakakis, Yannis. 1999. *Lacan and the Political*. London: Routledge.

Watson, James. 1968. *The Double Helix: A Personal Account of the Discovery of the Structure of DNA*. Harmondsworth: Penguin Books.

Wilson, Elizabeth. 2009. "'Would I Had Him with Me Always': Affects of Longing in Early Artificial Intelligence." *Isis* 100 (4):839–847.

Žižek, Slavoj. 1997. *The Plague of Fantasies*. London: Verso.

Žižek, Slavoj. 2006. *The Parallax View*. Cambridge, Massachusetts: The MIT Press.

7

THE ULTRA-CONTEMPORARY SELF AND HYPER-REDUCTIONISM OF HUMAN GENOME SCIENCE

Craig Venter

The epic moment that crucially shaped the subsequent course of the Human Genome Project (HGP) unfolded on 8 May 1998 at the business lounge of Dulles airport when Craig Venter, widely known as a maverick but brilliant scientist working with the private company TIGR (The Institute for Genomic Research), reached out to Francis Collins, director and head of the government-run human genome project, and declared that he aimed to complete sequencing the human genome by 2001, four years ahead of the schedule set by the public genome project. Mildly put, Collins described this news as both "offensive and preposterous". But the real feeling of "being kicked hard from behind" arrived a minute later when Venter declared that he wanted to coordinate and collaborate his efforts with the government-run project and then casually proposed, "so while we do the human genome, you can do mouse."[1] And that is how the human genome war was started – by Craig Venter.

On his own frequent admission, the human genome project for Venter was first and foremost about "winning" and "speed" and only secondarily about science and making money.[2] "If the Human Genome Project's byword was quality, Venter made it clear right away that his was speed" says historian of *The Genome War* James Shreeve (2004, 49). The matter of speed was as much about winning, it was also about claiming that he was better than "others". "We are going to be on the forefront of everything. We are going to need to build the faster computer in the world, with data production orders of magnitude bigger than anything else", he told Shreeve (2004, 3). The Nobel Prize–winning biologist Smith Hamilton, who was the principal geneticist on Venter's team, once described Venter in these terms: "Craig does not just want to be first, he wants to be first and the quality that will last for ages. He wants to stick it to the government program. Make them hurt" (Shreeve 2004, 311). In his autobiography, Venter time and again makes attempts to ascribe human genome sequencing as a matter of public good that is as important for human history as putting a man on the moon. He also describes this self-declared race as a battle for ideologies, morals, and ethics; however, it is impossible to ignore the fact that based on the kind of hyper-reductionist scientific choices he made in the process, human genome sequencing for Venter was less about humanity and more about the race to complete it first. In fact, Shreeve argues that without Venter there would have been no race to sequence the human genome, but more importantly without the race for the human genome, it would have been hard for Venter to define who he was (Shreeve 2004, 117). This chapter aims to discuss how Venter's affective self was isomorphic or consistent with his cognitive choices and how these choices also fundamentally defined the choices made by the "rival" side and thereby decisively shaped the hyper-reductionism of human genome science.

Mapping versus sequencing the human genome

Venter claims that his own choice became clear at 38,000 feet over the Pacific Ocean on the flight from Japan to back home in the United States in the form of an epiphany (Venter 2007, 121). By then, in 1990, Venter had spent ten years in finding and sequencing a gene of little more than 1,000 base pairs that coded an adrenaline receptor in the human brain. The only gene mapping and sequencing method available at that time involved painstakingly developing complementary DNA (cDNA) libraries of the messenger RNA likely to be responsible for coding the receptor protein.[3] The method that Venter followed involved surveying more than one million cDNA clones developed from genetic material from human brains to find one mRNA that contained the message to make the adrenaline receptor. When mRNA is isolated to make cDNA, a condensed version of protein-coding gene is made accessible in easy-to-read form, which then is multiplied by inserting it into bacterial plasmids. These cDNA were then sequenced by, what Venter described as a slow, manual method, known as slab-gel or gel-electrophoresis method.[4] This method of sequencing and mapping the needed mRNA was so tedious that Venter described it as the equivalent of finding a needle in a haystack – only monks tortured themselves with such tedious tasks (Venter 2007, 90). According to his own self-description as someone who was "addicted to speed" and "addicted to winning" spending dedicated long hours to locate some 100,000 estimated such needles in the haystack of the human genome was not Venter's style (Venter 2007, 17, 91).

This was also the time, between 1985 and 1990, the government run project of mapping and sequencing the whole human genome was being hatched in numerous meetings and seminars and symposia. The issue that fundamentally shaped these early debates was: Should the human genome project focus on finding and locating actual disease genes on chromosomes, called genome mapping, or should it focus on sequencing the entire three billion base pairs of the genome and postpone the gene hunting until after the sequencing? Behind this alluring question was lurking another: Was the DNA that did not code for protein – which at that time was called junk DNA – worth sequencing? And should not the project focus only on sequencing 1 to 3% of coding DNA – which clearly would require mapping first?[5]

At the time of early human genome debates in the late 1980s and early 1990s, the gene mapping followed the approach similar to Venter's own slow and methodical search for the adrenaline receptor in the human brain. It made sense to focus on finding genes first because the technique for mapping was known and tried and the technique for sequencing was so underdeveloped that it was predicted that the sequencing of the whole human genome could possibly be laborious and also staggeringly expensive (Shreeve 2004, 41). For the first five years of the project it was decided to "map first, sequence later" for these practical reasons. However, it wasn't just about the money and method either. There was an ethical issue involved in the whole purpose of doing human genome science. Early on, great doubts were raised against the complete sequencing of the human genome arguing that the sequence in itself without complementary information on mapping, without locating the sequences on the genome, would have been uninterpretable and hence worthless. The whole genome linkage map locating genes with respect to each other and deciphering relationships between genes and proteins and their function was instead proposed time and again as a slow, methodical, and conservative approach to the human genome project which would have relevance for therapeutic medical science.[6]

For Venter, the key to solving this foundational problem of the choice between mapping and sequencing was not only about finding and institutionalizing an effective and inexpensive method, but a matter of speed. This was a race to win. He indignantly declared in 1990 just about the time of the epiphany, "as I was to learn over and over again, a surprising number of human geneticists are often more concerned that they win the race to discover a disease gene than the race be finished as quickly as possible" (Venter 2007, 118). From the start Venter's goal was to establish rapid DNA sequencing as a valid approach for the human genome project.

The epiphany in the mid-sky happened in this context. As he claims, suspended in the sky, Venter came up with an approach to sequence randomly selected cDNA clones as a surrogate for the actual gene. At the time of the epiphany, however, the challenge was to obtain and sequence an entire gene, so-called full-length cDNAs. However, it is tough to obtain a full-length cDNA by applying "reverse trascriptase" enzyme to mRNA, because mRNA had a tendency to break in smaller fragments and hence could not be converted into full-length cDNA. Over a 12-hour flight, Venter pondered upon a question: "What if I just randomly picked a cDNA clone and sequenced it in one go?" (Venter 2007, 121). He thought about a somewhat reverse process, even if a cDNA was picked up randomly, it nevertheless was obtained from mRNA and hence was part of a real and expressed gene. This wasn't the only shortcut Venter thought of to address one of the foundational questions of the human genome project, i.e., the relative importance of mapping versus sequencing. His approach even short-circuited the second foundational dilemma, namely, whether the project should focus only on coding DNA or should even junk DNA also be sequenced.[7] Venter excitedly predicted that 1,000 randomly selected and sequenced cDNA clones would lead him to discover hundreds of genes. Because cDNA clones were not entire genes, just part of the gene sequence, they were nicknamed "Expressed Sequence Tags" (ESTs). Venter's team declared this approach as "the *fast* approach to facilitate tagging of most human genes in a few years at a fraction of the cost (italics mine)". This approach was even hailed at that time as an alternative to complete genome sequencing (Adams et al. 1991, 1661). ESTs approach to sequencing human genome was much criticized. ESTs are only a few hundred bases long and hence whether or not they legitimately represent genes was always a question. Many critics thought that ESTs could hardly tell anything worthwhile about the way the gene was expressed and the way it functioned (Davis et al. 2005). But at the time of his epiphany Venter was supremely confident that "sequencing a random selection of cDNA clones was going to be a very big winner" (Venter 2007, 123).

Venter's "winning" epiphany spread into yet another "winning strategy" and became a major push to finishing the human genome project faster than it was planned by the public laboratories. Venter and his team adopted and tested a method called whole-genome shotgun sequencing that challenged the methodical clone-by-clone approach of the public program. In the shotgun method, the entire DNA of an organism was shattered into tens of thousands of short stretches. Each piece was then sequenced by the automatic sequencing machines newly introduced in the market at that time and then these pieces were assembled together with the help of advanced software. The clone-by-clone approach adopted by the public consortium, on the other hand, involved sequencing clones selected from the specific region on the chromosome – they are that way already mapped and located on the genome. With the whole-genome shotgun method, Venter had thus developed another genomic shortcut to a faster route to the whole genome sequencing that involved no mapping at all.

By spring 1998, at the time of the epic moment of suggesting to the public human genome project "you do mouse", Venter had tried this method on two microorganisms and on baker's yeast before applying it to the human genome (Venter 2007, 123). In May 1998, soon after making the "mouse" suggestion to Francis Collins, Venter made a dramatic announcement at the 11th annual genomics conference at Cold Spring Harbor that his for-profit company (which was later named Celera, which meant speed) would complete sequencing the human genome four years earlier than the public project's deadline of 2005. The work thus projected to be completed in this deadline was stupendous.[8] Venter's announcement intensified the competition between public and private "racers". In his autobiography, Venter repeatedly called the five centres all over the world where the public-domain genome was being sequenced as "my self-proclaimed rivals" (Venter 2007, 287). In response to Venter's dramatic announcement at the Cold Spring Harbor conference, the British and American commitment to the public project was also doubled (McElheny 2010, 125).

As the storyline proceeds, soon after Venter's dramatic announcement at the Cold Spring Harbor conference, the public project also decided to abandon the "map first, sequence later" approach and adopt the whole genome sequencing (Shreeve 2004, 53). They retained their clone-to-clone sequencing method, which partially included mapping, because clones were located on the chromosomes, but they also bought the same automated sequencing machines as Venter's private company Celera did. It was first decided that if Venter was going to do the complete genome sequence by 2001, then the public program had to get the "draft" version out by then as well, even if it was full of gaps (Shreeve 2004, 124). Later it got even worse. The decision makers of the public project went a step ahead and declared they would complete the "draft" – not the finished genome as originally planned, but a "working draft" – by 2000 instead of 2001 – a year earlier than that of Venter's bait (Shreeve 2004, 191). In response Venter abandoned his own "high precision method" and decided to include the daily release of completed sequences by the public project into his own working draft genome.[9] After three years of intense competition between the public and private actors to arrive first at the finish line, as an anti-climax, eventually, a joint statement of the completion of the human genome sequencing was prepared and announced at the White House on 20 June 2000. However, at the time of the announcement, the so-called working draft was barely finished on both sides. The history of the human genome project, however, is never complete without the role played by Venter whose affective persona was described by one of the close observers as "drivingly ambitious, combative and full of himself" (McElheny 2010, 96).

Hyper-contemporary personality

The isomorphism between the cognitive and affective regimes in the human genome project was not just a prerequisite for the making of science, it actually shaped the very output of the scientific enterprise. What is interesting is to understand the precise historical form the isomorphism between affective and the cognitive have taken in the formation and completion of the human genome project. In the remainder of the chapter I want to locate the isomorphism between Venter's affective persona and his science within the framework of *changing psychology of our times* as discussed by the French philosopher Marcel Gauchet (Gauchet 2000, 2002). I want to show how Venter's hyper-competitive personality, especially the sentiment of "winning" as the core motivation for doing science, is not just a personal characteristic, but is symptomatic of the changing relationship between the self and the other in the contemporary historical times. Venter's hyper-competitiveness further signified the shifting connection between the individual and collective and is rooted in the changing patterns of socialization and the role of family and state in shaping these patterns.

Craig Venter can be characterized as what Gauchet calls a hyper-contemporary personality whose emergence, Gauchet theorizes, is marked by the pacification of conflicts, both political and with the self. This, Gauchet relates to the disappearance of what he calls the *revolutionary project* or (radical) *ideologies' mobilizing power*, and also the unprecedented rise of the welfare state in the post–World War II period which has significantly made other forms of collectivities redundant. Gauchet thinks that the pacification of conflicts goes hand in hand with the hyper-atomization of individual lives. Gauchet contrasts the hyper-contemporary personality with the ideological self of the previous era. This self was formed around identification with the social whole and rebellion against authority. The highest ideal of this self was heroism for the social good of one or the other kind. In contrast, the hyper-contemporary self is hyper-individualised, the basis of her being is not organized around embeddedness in any collectivity. This self is *disconnected* and *disengaged*.[10]

For Guachet, the end of a revolutionary project not only denotes a significant change in our ways of relating to ourselves and others, but it also implies a shift in the individual's relationship with the collective, and in socialization processes and the role played by state, school, and family in these processes. Gauchet argues that "socialization as adaptive learning consists of the incorporation of habits and rules that make collective coexistence possible . . . it is the process through which one learns to perceive oneself as somebody among others" (Gauchet 2000, 31). Gauchet calls this "training in anonymity" – "learning to abstract from oneself that sensitizes one to the collectivity, to the objectivity and to universality" (Gauchet 2000, 32). Gauchet argues that this radical distancing, decentring, and detachment from the self that enables an individual to learn to perceive oneself as somebody among others is called into question in the formation of the ultra-contemporary personality. *Adhesion to oneself* is the formative characteristic of an ultra-contemporary personality. Gauchet further argues that the contemporary period marks the shift towards an individualism of *disconnection* and *disengagement*, which means that "the demands for authenticity become incompatible with the attachment to collectivity" (Gauchet 2000, 32).

Gauchet further relates the emergence of hyper-contemporary personality with two broader changes in the role of state and family in socialization processes. In several of his publications, Gauchet argues that it is the monopoly attained by state in establishing and maintaining social bonds that has liberated the individual from having to think about being in society. There are two concomitant material changes at the root of contemporary individualism – first, the exceptionally sustained economic growth of Western capitalist societies since World War II that has created unprecedented prosperity and wealth, and, second, the development of the welfare state and the unprecedented rise of social protection.[11] For Gauchet this also coincides with the transformation of all forms of collectivity that transcend the individual, which Gauchet posits as the decline of the *mobilizing role of the ideologies* (Braeckman 2008).

Loss of collective transcendence/end of "revolutionary" project

In her book *Genes, Cells and Brains*, the science historian Hilary Rose reminisces in autobiographical style how the New Left, Marxists, and feminists critiqued the role of emerging biological sciences in the 1960s and 1970s. This critique happened in the background of the political ideal of radical, if not revolutionary, transformation of society and science. Rose, while discussing the human genome project, argues how this radical critique of science in the 1970s disappeared in the last 30 to 40 years. Rose further shows how during this period, the values of life scientists have changed from being "disinterested" in a Mertonian way to being "interested" in fame and fortune that commercial prospects of science have made possible by the emergence of the industrial-university-military complex (Rose and Rose 2012, 23). A similar trend is differently and widely acknowledged in STS (Science and Technology Studies) as a shift from mode 1 to mode 2 science. In mode 2, science has left its Mertonian norms of purity and disinterestedness and has become problem-oriented and ideologically and politically interested in the outcome and performance. However, mode 2 science is carried out in the context of application, it is generated in a wide variety of institutional spaces, not just in universities and industry. And, participants of mode 2 science have become more interested in the outcome of their science (Nowotny, Scott, and Gibbons 2001). This literature on the shift in science-society relationship points out that the political and social legitimacy of science is increasingly hinged on real or hyped assessment of consequences, performativity, applications, end results, and products, and not on radical critique of science's role in society. The overall emphasis on performance as science's legitimacy is also reflected in the public deliberation on science and technology. The knowledge that becomes controversial and highly contested – Funtowicz and Ravetz call this post-normal science (Funtowicz and Ravetz 1992) – is more than often about risk and uncertainty of consequences and performances and not about the political or philosophical critique of the front-end science.

According to Rose's biographical account, the radical political project of democratizing science came in three waves. The first was the social relations of the science movement that emerged in the Cold War period, also partly in response to Lysenko's fraudulent science. The second wave was the radical science movement of the 1960s and 1970s calling for larger societal accountability of science. And the third wave began in the late 1970s and early 1980s with the German Greens leading the opposition to geneticization of biomedicine and also genetic engineering and GM crops (Rose and Rose 2012, 34–36). Rose discusses how biologists were morally outraged with the way their science on plant hormones was abused by military scientists to produce chemical weapons that destroyed forests and crops and people in Vietnam. In the 1960s, along with the global opposition to the Vietnam War, biologists' outrage was expressed in journals, teachings, and demonstrations on the streets from which, according to Rose, was born a new radical science movement (Rose and Rose 2012, 14). I have no doubt that there are multiple other ways to construct the history of a radical science movement. I have mentioned Rose's partly autobiographical account, she being a renowned historian of life-science herself, as one possible way to understand the end of ideological or radical critique of science and society at the time of the human genome project.

Gauchet also remarks about this ideological era when the idea of the individual embedded in society was integral to the radical project of critique of society. The radical project also had logic of individualization but this logic began with the general and proceeded to particularize it (Gauchet 2000, 27). Venter's combative affective self at the core of his science had no association with the radical project of the critique or transformation of science and society. His logic was inversion of the radical logic, i.e., Venter began with the singular "being-with-the-self" and went on to universalize it by forcing the human genome project to change its direction and focus. Although Venter frequently acknowledges in his biography how his involvement in the Vietnam War "compelled him to understand life in its most intimate detail", and how it made him know the "fragility of life" (Venter 2007, 2, 26). This revelation, however, was more of a personal and existential nature than political, collective, or societal in any way. At the young age of 17, Venter received his draft notice in 1965 to serve in the Vietnam War and although he was personally against the war, and consciously declared himself as a pacifist later, he was inspired to go to Vietnam by his family history of military service. After 14 months of medical training in a Navy hospital, he was sent off to Vietnam in 1967 (Venter 2007, 20). Vietnam remained a pioneering experience in his life, one that he repeatedly recalled later in his life. It is surprising that he did not mention this in his autobiography, but he confided in James Shreeve in 2002, after the induced genome race was over, that while serving as a medic in Vietnam he tried to kill himself. Mending to the tangled bodies in the war zone had made him realize not only the purposelessness of the war but also the meaninglessness of life itself. Venter described the attempt to kill himself as a turning point in his life (Shreeve 2004, 359). Even at the pinnacle moment of fame in his life, at the historical moment of joint announcement of the completion of the human genome sequencing at the White House in June 2000, while the world was watching, he was worried that reminiscing about the Vietnam experience would make him emotional in front of international media. In his speech he acknowledged how far he had come since his period in Vietnam and how his time in service had given him drive and determination and how this experience taught him the fragility and importance of human life and spirit (Venter 2007, 313). Venter's association with the Vietnam War was thus deeply existential and personal and far from political.

Venter's hubris, his desire to be known as a winner, turns out to be incompatible with other collective considerations. He was hardly bothered about the relative merit of mapping against sequencing. The motivating drive in his life was to prove that he was "authentically" the best; his deeper affective needs were to *adhere to oneself*, and to *detach* and *disconnect* from any other form of collective – political or ethical – considerations. Venter's team was also entirely mobilized to work hard, almost insane hours, on the motive of "doing it fast" and "winning". In fact, when in 2000 Francis Collins and Craig Venter mutually decided to declare the "truce" and agreed to make a joint announcement of completing the genome at the White House, all the lead members of Venter's team were upset when they came to know about Venter's decision to declare that the race ended in a tie (Shreeve 2004, 345–346). In fact, both sides decided to call it a tie only when the chances of a clear win on either side were slim. It was not in any remote sense a moral, ethical, or political consideration that made this decision possible. Venter's combative personality, the motivating framework of winning and speed, and their imposing influence on the entire course of the human genome project must be understood in the context of the erosion of, what Gauchet calls, the radical ideologies' mobilizing power, the resulting disenchantment of the political, and the disappearance of the viewpoint of not just the political but also the ethical. This is the context in which Gauchet argues how in our contemporary times "the individual has overtly come to the fore" (Braeckman 2008, 31).

Culture of narcissism/competitive individualism

There is a paradox, however, in the decline of the collective and the rise of the "individual". This self is also overtly dependent upon the approval of "others". This is not just Hegelian insight that the self occurs in relation to others, or, in other words, the self can be truly realized only in relation to the other.[12] Gauchet in fact argues that the pacification of conflicts with the self and with the other is at the heart of the hyper-contemporary personality. Gauchet relates the pacification of the conflicts with the end of the era of repression – marked by the disappearance of the authoritarian style of education, including sexual opening up and the emergence of hedonistic culture. With the pacification of conflicts, being oneself no longer means being-with-oneself, i.e., knowing who one is, but being-with-others and being-with-neighbouring-world. Gauchet puts this changing relationship with the self and other in these words, "the essential no longer concerns the relation of the self to the self but the relation of the self to a remaining semi-obscurity". This self does not possess itself but flows in the "universe of networks" (Gauchet 2000, 37). Gauchet does not mean to argue that the divisions or contradictions of desire have disappeared or that the human desire is no longer structurally divided, but that the violent and intense conflicts these divisions gave expression to have pacified. Gauchet, however, counts inner emptiness and rage; annihilating solitude; even fear of others; the other as threat; and distance, distrust, and avoidance as defining aspects of the ultra-contemporary personality. Most importantly, "perception of others as intrinsically threatening" is the principal feature of hyper-contemporary mentality. "The ultracontemporary individual not only flees others, all the while fearing to lose them, he is also in flight from himself" (Gauchet 2000, 37–41).

Venter described his childhood to James Shreeve as "difficult" precisely in this sense of the crisis of identity pervasive in a hyper-contemporary personality. As told to Shreeve, Venter as a 16-year-old boy did not "give a shit about going to college". He was a bad student by choice, "proactively". He told Shreeve that he failed tests and kept exam papers blank even if he knew the answers. He masked his rejection of the world with contempt (Shreeve 2004, 68). Based on Venter's account, Shreeve argues that in comparison to Venter's elder brother, who was a studious, brilliant student with unlimited prospects, Craig was "given" a role of bad egg in the family, was named as a "failure". Shreeve thinks this was implied in his absence from the recorded history, videos and photos, of the family (Shreeve 2004, 69). As per his own autobiographical narrative, early in life Venter felt that he could not be like his elder brother Gary – spectacularly success-ful. He therefore tried to get his parents' attention by failing spectacularly (Shreeve 2004, 54). In his own autobiography, however, Venter presents himself as someone who from an early age liked taking risks and facing challenges (Venter 2007, 5–9). The adult brother Gary thinks that he and Craig were not different in their behaviour when young, just that Craig wanted to get caught by the parents while pursuing a bad boy image (Ven-ter 2007, 11). As a boy Venter identified with the hero of J. D. Salinger's *The Catcher in the Rye* – a victim of institutional malaise who was rebel-lious, was unappreciated in school and family, and who in turn rejected both (Venter 2007, 15). Venter looks back in his autobiography at how his parents rewarded his elder brother for being a super-achiever and how he tried to get the attention of his parents by being a super-underachiever. As a boy, Venter finally found his niche by winning swimming races – also against his brother: "Winning was a new experience for me, and an addic-tive one" (Venter 2007, 17). But he carried this image of a bad boy, the image of super-underachiever, as a marker of his identity, his contemptuous rejection of authority, to his adult self. Shreeve reports that while still a young scientist-in-the-making, he would go to faculty meetings "full of bra-vado and blind confidence . . . would tell the distinguished professor types that they were stupid", but paradoxically, then, he would also feel pained wondering "why he wasn't accepted by the others" (Shreeve 2004, 75). On his own admission, while being in tussle with the NIH for funding at the beginning of his career, he declared, "determined not to waste my life, I reject absolute authority unless it is rational" (McElheny 2010, 95). After the May 1998 announcement that he would complete the sequencing of the human genome several years in advance of the public project's finish line, he made a lot of new enemies, while the old ones "galvanized in uniform opposition" (Shreeve 2004, 8).

All these aspects of his personality are often interpreted as hubris, as traits of intense competitiveness, as eccentricities of a maverick scientist. But there could be another interpretation. As a child and throughout his adult life, Venter projected a personality that apparently created an image as if he did not want others' approval. "Not being like" others was formatively driving his personality. However, precisely by defining his identity in terms of "not being like" and "not wanting others' approval" his entire existence and persona in the world was defined by the "others". He needed others, their disapproval, opposition, their rejection and antagonism, to know who he was. His own inner psychic orientation in the world was abysmally weak in the absence of this hate-reject-win relationship with others. Throughout his life, Venter didn't display any strong adherence to any political, ideological, aesthetic, or spiritual belief. He did not identify in life-defining ways with a cultural model or a person, parent, or a teacher, or with a spiritual, scientific, or political idol. Gauchet argues how identification with such models provides a world view of human life, and through them one enters into communication with the ideal. It is hard to find such strong communication with any ideal in Venter's life. Gauchet calls this "weakness of identification" or "de-identification" because there is "de-idealisation" (Gauchet 2000, 39). However, as Gauchet further argues, in the case of the hyper-contemporary individual there is something deeply democratic about the rejection of authority – "the person constructed this way will not follow the *Führer*" (Gauchet 2000, 39). But, the lack of identification with the authority of ideals, ideologies or role models creates a personality that suffers from the crisis of consistency, the lack of it (Gauchet 2000, 39). "Not being like" peers as others is the only source that defined Venter's place in the world. Venter's aggressive competitiveness and his hubris as the only means to relate to the professional peers was an expression of this trait of hyper-contemporary personality. However, the underlying message of the hubris and competitiveness, "I am not like you", and "I am better than you", meant that he "needed" others in a life-defining way. The same others were feared at the same time because losing them would be life threatening (Gauchet 2000, 35). This self-other dyad marks the profound uncertainty about identity and equally profound failure to constructively master relations with "needed" but "feared" others in social life (Moyn 2009, 338).

Hyper-contemporary culture

I must clarify that not all principal actors in the conundrum of the human genome project displayed such straightforward traits of hyper-contemporary personality. As a stark contrast to the hyper-contemporary personality of Craig Venter, I want to briefly discuss the affective traits of another principal actor of the human genome project to show how despite these fundamental affective differences the human genome project in the end was decisively driven by the choices made by the hyper-contemporary affective self – i.e., Venter's. Not only the fact that all other actors justified and followed Venter's choice of the finish line but also his choice of method, i.e., sequencing instead of mapping, ultimately demonstrated the cultural hold of hyper-contemporary psychology in our times.

John Sulston was the director of the Sanger Centre, privately funded by the Wellcome Trust. The centre was part of the international team of the public human genome project – one of Venter's self-proclaimed rivals. Sulston was jointly awarded the Nobel Prize in 2002 for his work of sequencing the whole genome of nematode *Caenorhabditis elegans*. Sulston also wrote his autobiographical account of the human genome project *The Common Thread* which was published in 2002 (Sulston and Ferry 2002). He entirely frames his account as a tale of personal failure and a moral battle. There are two actors in his story, John Sulston as a protagonist and Craig Venter as an antagonist. In fact, Sulston in his own description of the self emerges as an antithesis of Venter. It is a tale of how a somewhat withdrawn scientist interested only in his own small world of research was transformed into a reluctant leader of an international project on the human genome. He describes how in the course of the human genome project not only how his small-scale research practices turned into big science endeavours, but how his persona transformed from being an indiscreet, small-time researcher into a visible, (minor) celebrity scientist (Sulston and Ferry 2002, 212). In complete contradistinction of Venter's confident writing style, page after page declaring how he developed "winning strategies" and how he is going to "hurt others by winning the race", and how he was better than others, the underlying tone in Sulston's autobiography is that of a failure – he tells a tale of how he could not seize political opportunities of funding and how at crucial junctures he could not act in time because of his own reluctance and insecurity. Sulston thought that the basic flaw in his personality was a certain lack of "megalomania" (Sulston and Ferry 2002, 134–137). Sulston also regrets how because of his reluctance, insecurity, and lack of certain ability to imagine things in larger than life terms he personally lost the "race" to other actors in the public side of the human genome project. But the most interesting part is how in more than one way Sulston has framed his autobiography as a story of a moral crusade against Venter. Early on he declares that he finds the metaphor of race for the scientific quest misleading and even disgusting (Sulston and Ferry 2002, 7). Throughout the early history of the project, he studiously advocated mapping and not sequencing as the mainstay of the project. But the main story of the moral crusade is weaved around the question of patenting and the free access to genome data emerging from the project. He declares the central aim of the human genome project was mapping (and even sequencing) the genome for the free access to scientific community. He also declared that the founding of Venter's commercial company Celera Genomics was a threat to the future of biology. He takes pride in stating the importance of his personal role in the battle over "access to the most fundamental information about humanity". "Today any scientist anywhere can access the sequence freely at no cost and use the information to make his or her own discoveries" (Sulston and Ferry 2002, 8). He then declares that he (we) wrote this book (his autobiographical account of the Human Genome project is co-authored) "so that people might understand how close the world came to losing that freedom" (Sulston and Ferry 2002, 8). He writes the book as a moral tale of a battle between those who wanted to keep the information about the human genome freely available to the research community as a common heritage to humankind and those who wanted to privatize it for personal gain. That is, a battle between him and Venter. Even on the face of the doubts expressed by other actors from the public consortium, Sulston stood his ground on the issue of basic ethics of science, and proactively wrote the "Bermuda Principles" that made it mandatory for all public human genome partners to release their sequencing data every day in the public domain (Sulston and Ferry 2002, 168). Even after the completion of the human genome project, and especially after he was awarded the Nobel Prize in 2002 (not for his work on the human genome, though), Sulston retained his staunch commitment to the campaign opposing patenting of the genome. He is also one of the 22 Nobel laureates who signed the humanist manifesto in 2003.

Clearly, Sulston shows many personality traits that are contrasting to that of Venter's. He does not look as dependent upon others' (dis)approval as Venter is for his psychic existence. Having shown such clear devotion to preserve the collective interests on the question of patenting, Sulston's individuality is not defined by the detachment and disengagement from the perspective of the whole. Implied in his ethical choices with respect to the human genome project and also in his endorsement of the humanist manifesto is an affective life that is compellingly driven by the communication with the aspired ideal world.

Despite these crucial differences, the curious part of the story is that Sulston failed to have a decisive influence on the induced "race" between the public and private initiatives and hence the direction of the human genome project. In fact, at crucial moments he not only gave up his own strong beliefs but even justified the one propelled by Venter as the right one. At the time when in May 1998 Venter declared his intentions to complete the sequencing four years ahead of the public project finish line, the public consortium had a chance to not take the bait and to stick to its own strong preference for mapping instead of sequencing and also completing the finished draft by 2005 as it was originally decided instead of producing a working draft by 2001. Soon after Venter's declaration, in the debates that followed internally among the main actors of the public consortium, Sulston's was a voice that commanded the decisive moral authority. Soon after Venter's public challenge, Sulston and his close colleague Robert Waterston from Washington wrote an article in *Science* in which they first emphasized the public consortium's ultimate commitment to produce "a comprehensive human genetic map" on which, as speculated at that time, the position of 30,000 genes were located. They also declared that since 1990 the human genome project had identified the position of 200 such disease-associated genes by the method that involved both mapping and sequencing by adopting the clone-by-clone approach. According to this method the clones of the size of 400 to 1,000 base pairs representing the region being studied on a chromosome were selected and sequenced by the gel-based separation method and the reads were assembled together based on the sequence overlaps of the clones (Waterston and Sulston 1998, 53). The authors then criticized and declared the whole-genome shotgun method, which entailed subcloning and sequencing random fragments of total human genomic DNA, even with the tenfold (10x) coverage of the genome, as "woefully inadequate".[13] They showed that this was chiefly because of the repeats – because 40% of the human genome was repetitive and many repeats exceeded the size of a single sequence read making them impossible to locate on the genome in the absence of the mapping of the random fragments on the genome in the first place (Waterston and Sulston 1998, 54). In this article Sulston and Waterston reiterated their ultimate commitment to producing the genome of high quality – the error rate of 1 in 10,000 bases – by the originally set date of 2005. However, curiously, they also in the end approved generating a one-third completed working draft by 2001, also by using automatic sequencing machines to speed up the sequencing process. What they did not compromise, even on the face of the high pressure from other colleagues like Eric Lander of the public consortium, was their view that the sequencing made sense only when the reads were mapped on the genome and thereby retained their strong preference for sequencing of only the mapped clones (Waterston and Sulston 1998). Sulston was clearly able to exert his moral authority on the choice of the method of the public project, i.e., sequence only mapped clones, however, at the crucial moment he gave up his moral and even material authority to influence the "race" between the private and public actors and thereby influence the quality of the output coming from both sides. When Francis Collins accelerated the race by proposing a new finish line – to complete a working draft in spring 2000, a year in advance of the new finishing line set by Venter in 2001 – Sulston conceded by saying it was the right thing to do (Sulston and Ferry 2002). Collins was most worried about Sulston, because if Sulston had decided to go his own way at this crucial moment, Collins (and even Sulston) knew that the Wellcome Trust would have also withdrawn its support of the public consortium and would have proudly backed Sulston only. Both Sulston (and also Collins) were aware that he had a decisive moral say in the process. By supporting Collins' proposal to accelerate the race, which otherwise was absolutely unnecessary, at this decisive moment, Sulston's need for others' approval had an upper hand to his inner psychic gyroscope that made him psychically independent of others.

In my interpretation, the idea that the race found wide acceptance among the diverse set of actors of the human genome project is a testimony to the fact that the hyper-contemporary psychology of our times had dominant cultural sanction, despite the fact that several individuals themselves may not entirely display such affective traits in their personalities as Venter does. Venter's decisive role in defining the core affective motive of the project was a triumph of these cultural traits. This triumph was not just one individual's competitive power over others.

Hyper-reductionism of science of the human genome

The human genome war eventually resulted into a form of hyper-reductionism. Ultimately the preference for sequencing reduced the notion of the gene as merely sequences of nucleotides outside of any framework in which the gene expressed itself. Even when the actors of public consortium took pride in claiming that their sequenced BACs (clones) were already mapped on the chromosome, and that their sequencing was systemic and their individual clones were sequenced to 4x (fourfold) coverage, as

Sulston claimed (Sulston and Ferry 2002, 214), but in terms of gene expression and functioning and its role in cellular operation, the public genome was as unintelligible as the one done by Celera Genomics. Not only did the war result in such hyper-reductionism, but the science of sequencing itself was arguably of dubious quality on both sides. John Sulston, apropos to his moral ability to own up to it, said in his autobiography that at the time of the announcement at the White House on 20 June 2000 the public consortium had not got to its promised 90% sequencing mark and even Celera's data were known to be thin and no one was really ready to announce, but "we just put together what we did have and wrapped it up in a nice way, and said it was done. . . . Yes, we were just a bunch of phonies!" (Sulston and Ferry 2002, 224). By the time two genomes were published seven months after the White House announcement, some people on the public project side were not satisfied; a senior colleague from the Department of Energy, one of the principal members of the consortium, announced that "he was ashamed to be associated with it" (Shreeve 2004, 389). Phil Green, an ardent critic of Celera, also owned up to "gross inefficiencies in the process of sequence acquisition" on the public side (Green 2002, 4144). Shreeve absolves the Celera version of the genome by calling it as "more complete and usable" (Shreeve 2004, 367). Here, Shreeve's own bias as a historian is palpably visible also because he provides no evidence to substantiate his statement. A powerful challenge to Celera's method and approach came predictably from Sulston et al.

From the very beginning, serious doubts were raised on the whole genome shotgun method especially on the ground of what was called "the problem of repeats". Much before the debates on the human genome project started in the mid-1980s, it was known that a sizeable portion of the human genome is composed of repetitive DNA sequences. The sequences that appear to occur only once in the genome are called unique or single-copy sequences. As per one estimate made in 1981, 65% of the human genome was understood as composed of single-copy sequences and the remaining 35% as repetitive sequences interspersed with single-copy sequences (Tashima et al. 1981). Philip Green, a mathematician geneticist in the public genome project had early on pointed out that machine-read sequences of broken down random bits of genome cannot be assembled together unless they were already located and mapped on chromosomes. If the repeats were not located or mapped on chromosomes prior to sequencing, the assembled genome would be a jumble of misplaced bits. In fact, it was speculated that it would have been impossible to find out where the repeated sections of the genome belonged. Green's was a powerful critique of the whole-genome shotgun method. He especially pointed out that the computing simulation model adopted by Celera for the whole genome assembly after the sequencing was based on a wrong assumption that repeats were randomly distributed in the genome, whereas the repeat density varied widely, often occurring in non-random clusters (Green 1997, 413). When Francis Collins declared a new finish line, Venter in response decided to limit sequencing to only 4x and the rest use the government project data released nightly in the public domain for Celera's own assembly. This not only made Celera's sequencing faster but it also allegedly saved $100 million. However, the public project had an entirely different method of the clone-to-clone approach to sequencing that was not compatible with the whole-genome shotgun approach and eventually the mixing of these two desperate sets of data created massive computational problem for Celera resulting in equally massive compromises on the quality of the assembled genome – as later shown by Sulston et al. from the public consortium.

After both papers on the working draft of the human genome sequence – the one by the public consortium in *Nature* (Consortium 2001) and the second by Celera Genomics in *Science* (Venter et al. 2001) – were published, Sulston along with his colleagues Lander and Waterston published a critique of Celera's whole-genome shotgun method (Waterston, Lander, and Sulston 2002). The authors alleged that the Celera team "decomposed" or shredded publicly released data of the public consortium and combined and incorporated them with their own random fragment shotgun data. They argue that even a twofold coverage of clone-to-clone sequencing of public data would have been enough for recovering the entirety of the genome assembly because sequence-tagged site maps or BACs were already anchored or mapped to the genome. Based on a detailed computational analysis, they further trashed Celera's paper by arguing that the difference between the assembly put together by combining the public consortium data with the Celera shotgun data and the assembly made from the public data alone was slight. Sulston et al. clarified that they did not think that the whole-genome shotgun approach could not be successfully applied, but that the approach followed by the Celera paper did not provide such evidence. The extent of the coverage of the genome in Celera's draft was also a major issue of contention. Sulston et al. estimated only 2.9-fold coverage at one point and at another 1.75x instead of 10x as originally pledged and necessary (Waterston, Lander, and Sulston 2002). Another public consortium scientist, Phil Green, also contested if Celera's genome was original, especially when the BAC sequences from the public project were shredded and used in a way that all original information from the public sequences was retained and included in Celera's assembly. Green even accused Celera of using unshredded and assembled public BAC sequences to close the gaps. And despite inclusion of the public data, according to Green, 20% of the genome was missing in Celera's assembly because 116,000 small islands of sequences could not be placed on the finished assembly (Green 2002, 4143). Similarly, in the second paper replying to Celera's reply, Sulston et al. further alleged that approximately 60% of the sequence data and 100% of the mapping data used in Celera's analysis came from the public data; also the public data were used in a variety of ways that implicitly preserved the assembly of the public project. The authors concluded that "Celera's assemblies made extensive and inextricable use of the HGP genome information and this was not an independent assembly of the human genome" (Waterston, Lander, and Sulston 2003, 3022). The Celera team replied to Sulston et al.'s objections and the dialogue between two warring sides continued in the pages of *Science* for another two years. It is hard to take sides, but I have emphasized here that the public consortium's challenges to Celera's genome show that Celera's draft was not that complete and it was not independently produced, and its quality was not above contention, as Shreeve claims. The scientists from the public side were willing to own up to the questionable quality of their own draft in any case. Eventually Celera did not come up with the draft assembled entirely with the in-house shotgun data as it had repeatedly declared, but the public consortium did come up with the 99% finished version in 2004, claiming only 341 gaps and an error rate of 1 per 100,000 bases (Consortium 2004).

More than a decade later the results of a sequel to the human genome, called ENCODE (Encyclopedia of DNA Elements launched in 2003), was released in 2012 which was a combined work of 400 scientists from around the globe. The results of the projects are discussed at length in chapter 5. The ENCODE has thrown several more surprises, which cannot be discussed here, but the point is that it took a decade (and more) longer to achieve meaningful understanding on the nature and function of the genome. It is highly questionable if the induced "race" produced any meaningful results of the human genome project that took another decade to meaningfully decode.

Discussion: Who is the scientist-subject doing science?

In the recent texts in the history of science that look at subjectivity-objectivity relationship, the scientist-subject is more than often theorized as an autonomous, rational subject wilfully driven by social and scientific ethos. Alternatively, the scientist-subject is a Foucauldian construct – the subject reduced entirely to the disciplinary effects of power. The processes of subjectivity are, however, underdetermined and complex. As Jane Flax, a feminist scholar and psychoanalyst, describes, psychic structures are constituted by interweaving of many hetero-geneous structures and capacities. These include complex cluster of capabilities, modes of processing, altering and relating experi-ence, and foci of affects, somatic effects, and transformation of process into various kinds of language, fantasy, delusion, defenses, thought, and modes of relating to self and others.

(Flax 1993, 93–94)

Such a multiple, heterogeneous, contradictory – the feeling, suffering, experiencing – subject is rarely a part of this recent history of science. In a similar vein, Söderqvist discusses at length how the existential project of the individual scientist is not the subject of science biographies and how the inclusion of the scientist-subject doing science as making existential choices of self-assertion and self-renewal can provide a different concep-tion of science (Söderqvist 1996). In the history of science, even when the subjectivity is accounted as co-implicated with the objective science, it is treated only as the mental states of collectives.[14] Steven Shapin warns against the risk that understanding "actors' categories" as the goal for history of science can dissolve the subject-matter of science to the vision of individual scientist. And hence, he further suggests, the indi-vidual reflexes should be disciplined by sociologists' collectivism (Shapin 1991, 354–355). While the methodology of science as emotionally neu-tral is increasingly challenged, the real dividing lines are drawn on the question of whose emotions count – individual scientist's or collective's?

Firstly, this chapter shows how Venter's choice of the hyper-reductionist method of sequencing (without mapping) of the human genome was driven by a highly personal affective motive of winning and proving oneself bet-ter than others. This points out that the scientist-subject is not selecting his empirical method from an infinite list of possibilities (as if choosing a product in the supermarket) but he chooses the "necessities that deter-mine itself" (Žižek 2006, 79). The chapter thus questions the image of scientist-subject as neo-Kantian ideal – a rational and coherent entity mak-ing deliberate and calculated theory choices based on empirical evidences. It instead highlights the scientist-subject as motivated by deeply affective and even existential drives split between the conscious and unconscious.

Secondly, the chapter discusses how Venter's hyper-contemporary persona – winning and proving one's self as better than others – provided powerful motivational force for the choice of empirical method in the science of the human genome and how the making of self and science were profoundly constitutive of each other. In asking what was the emo-tional a priori of the scientist, this chapter presents an alternative history of reductionism in human genome science. Such an approach aspires to partially answer the question Keller asks: Why is it that the nature-nurture debate resists resolution and why does the reductionist language of the particulate gene persists despite contrary evidences for more than a century? (Keller 2010, 10–13). This chapter shows isomorphism between the affective and the cognitive by "seeing" the things that social history of reductionism misses.

Thirdly, in two significant ways this chapter transcends the polarizing dichotomy of the individual versus collective approach towards understanding the role of emotions in the history of science. These two approaches do not necessarily represent two completely independent concerns. The chapter shows how Venter's choice of method was not only compellingly related to motivations of his affective persona but how this choice decisively shaped the course of the entire project. It shows how the scientific ideas and institutions and the affective choices made by the individual scientist mutually reinforce each other. Furthermore, the chapter shows how the affective life of Venter is far from what Shapin called "atomizing particularism". It can be described in the way that combines what Söderqvist separately defines as three different approaches to scientific biography: social biography that focuses on scientist's situatedness in certain cultural and historical time; psychobiography that explains scientist's personality traits with reference to psychological theory; and Söderqvist's own approach of existential biography that foregrounds the practice of science as scientist's quest for self-assertion and self-realization (Söderqvist 1996). Venter's affective persona thus discussed in the chapter is not a collection of atomizing idiosyncrasies but grounded in the social and cultural history of collective emotions and his own developmental psycho-social history. The chapter discusses how Venter's affective self is the product of the shifting psychology of our times and how this individual psychology is manifested only in relation to the "other" and the collective.

Notes

1 As described in Shreeve (2004, 21–22).
2 Reference to this comes up several times in Shreeve's account of *The Genome War* (Shreeve 2004, 49, 101, 102, 117, 311).
3 The messenger RNA molecule abridged the genetic code and the protein. The working or functional copy of the human DNA could be spread over as many as a million base pairs but the edited messenger RNA that actually codes amino acids of protein could be only a thousand base pairs in length. But messenger RNA is transitory and unstable. By using a restriction enzyme, especially the one called "reverse trascriptase", RNA can be converted into a stable form of DNA which is called complementary DNA or cDNA.

4 This sequencing method originally pioneered by the Nobel Prize–winning scientist Fred Sanger was later modified by Walter Gilbert by including radioactive labels to identify and sequence four nucleotides. For the description of the method, see Maxam and Gilbert (1977). Also for a fascinating history of the method as it was being developed, see Sanger's autobiographical account (Sanger 1988). The method involved loading cDNA fragments of the length 600 to 1,000 base pairs long onto one end of a 30-centimetre-long polyacrylamide gel sandwiched between glass slabs. The DNA is then made to migrate through the gel by applying a low amount of electric charge. Because smaller fragments travel through the gel faster, fragments of different sizes gradually separate. This method is slow, takes about four hours for the loaded fragments to travel through the gel. And the process has to be repeated many times before a reasonably reliable sequence of the cDNA clone is available. Also, it worked best for the fragments shorter than 600 base pairs. For the detailed description and history of the method of gel electrophoresis, see Chiang (2009). For the review of the method, see Swerdlow and Gesteland (1990).

5 Lewin reviews one of these early meetings at Cold Spring Harbor on the merits of human genome sequencing versus mapping and reports how several scientists "shuddered" at the idea of sequencing and how it was considered premature (Lewin 1986a). Lewin reports another such meeting in which the dominant mood was to prefer mapping to sequencing (Lewin 1986b). A rejoinder was later published in *Science* with an argument that sequencing should not be discounted only on the basis of costs and that the cost of sequencing is likely to dramatically reduce over the years (Smith and Sinsheimer 1986). For another rejoinder on the relative demerit of sequencing the entire genome that contains a substantial amount of non-coding regions, see Gall (1986). The debate continued when even a decade later the challenges of sequencing were still being discussed (Rowen, Mahiras, and Hood 1997). In the meanwhile, the proposal to carefully combine mapping and sequencing was also discussed (Olson 1993). For the review of the science of physical mapping of the human genome, see Olson et al. (1989). Francis Collins also reminisces about some of the early issues in his review article published in 2003 after the race for human genome was complete (Collins, Morgan, and Patrinos 2003). For further discussion on early debates in the history of the human genome project, see McElheny (2010, 62–70). The matter of relative merit of sequencing versus mapping has also been debated among the philosophers of molecular biology. Kitcher supported the idea of genetic mapping and subsequent sequencing, but philosophers Rosenberg and Sarkar raised serious doubts about the usefulness of the full genome sequence, which was considered wasteful, misguided, and even loaded with possibilities of social harm (Kitcher 1994; Lewontin 1992; Rosenberg 1994; Tauber and Sarkar 1993).

6 Throughout these early debates, several scientists at different points in time persistently used a number of metaphors to argue against just the sequencing of the genome. David Botstein, one of the most vocal scientists advocating gene mapping, declared that gene sequencing "would be like a complete set of Egyptian hieroglyphics with no accompanying text in a known language". David Botstein remained committed to finding appropriate methods to mapping the genome of diverse organisms that reveal the order and logic of a genetic

155

program rather than the physical location of genes on chromosomes. For the discussion of one such method and its relevance for mapping genes, see Brown and Botstein (1999). For the discussion on constructing genetic linkage maps of the human genome and relating the linkage maps with inherited traits, see Botstein et al. (1980).

7 Since the discovery in 1977, genes are understood as broken up into "expressed" pieces, i.e., coding regions known as exons, interspersed by non-coding regions called introns. When a protein is to be made, the cell makes a crude RNA copy of the entire coding region containing exons. So cDNA copies made from applying reverse transcriptase enzyme to RNA contains only exons, no intervening junk of introns. For the debates on anti-reductionism and the complex functioning of the genome, see Rosenberg (2006, 31).

8 Although the tone in the Venter et al. paper that followed the announcement in May was conciliatory – "we look forward to a mutually rewarding partnership between public and private institutions" – the paper also outlined the whole-genome shotgun method, introduced the functioning of automatic capillary sequencing machines, and compared the merits of the whole-genome shotgun method with the sequence-ready maps of BACs – all in perfectly mild and professionally academic tone (Venter et al. 1998).

9 Before the mass production of sequences began, the public project centres had come to an agreement that all sequences longer than 1,000 bases will be released in the public domain every day. This declaration became known as the "Bermuda Principles" which was made mandatory for all who sought a government grant for sequencing. This step was taken on the initiative of the British scientist John Sulston and his close American colleague Robert Waterston who strongly wanted to prevent private patenting of the public data. For the detailed account of the moral battle between Sulston and Venter, see Sulston and Ferry (2002).

10 For the detailed discussion of these arguments, see Gauchet (2000). Also, for the further discussion in English on Gauchet's original work in French on democracy and psychology of our times, see Braeckman (2008). For the redefinition of the unconscious, see Gauchet (2002).

11 Only a few of Gauchet's writings are translated from French to English and hence I had to unfortunately depend upon some secondary sources. For the summary in English of some of Gauchet's important arguments, see Braeckman (2008).

12 For the feminist and psychoanalytical deliberation on the Hegelian concept of recognition, see Benjamin (1988). Benjamin's work is also discussed in detail in Chapter 5.

13 The whole-genome shotgun method adopted by Venter had a problem: each sequenced DNA coming out of the automated machines was at the most 500 base pairs long. That means that, for instance, the genome of *Drosophila* would contain 240,000 pieces to be put together by advanced computing machines, but the sequencing in the shotgun method was so random that some parts would be represented many times over and some would not be represented at all. This effectively meant that the whole genome would have to be sequenced at least ten times over – or, 10x or tenfold times, to make sure the whole territory of the genome was sampled and sequenced.

14 Daston and Galison in their historical analysis of the co-construction of subjectivity-objectivity insist that they were concerned with the collectives

and not with the individual psychology (Daston and Galison 2007). A detailed review by the author of Daston and Galison's *Objectivity* and Daston's other work on the role of affect in scientific observation is discussed in Chapter 2.

References

Adams, Mark, Jenny Kelley, Jeannine Gocayne, Mark Dubnick, Mihael Polyme-ropoulos, Hong Xiao, Carl Merril, Andrew Wu, Bjorn Olde, Ruben Moreno, Anthony Kerlavage, Richard McCombie, and J. Craig Venter. 1991. "Complementary DNA Sequencing: Expressed Sequence Tags and Human Genome Project." *Science* 252 (5013):1651–1656.

Benjamin, Jessica. 1988. *The Bonds of Love: Psychoanalysis, Feminism, and the Problem of Domination.* New York: Pantheon Books.

Botstein, David, Raymond White, Mark Skolnick, and Ronald Davis. 1980. "Construction of a Genetic Linkage Map in Man Using Restriction Fragment Length Polymorphisms." *The American Society of Human Genetics* 32:314–331.

Braeckman, Antoon. 2008. "The Closing of the Civic Mind: Marcel Gauchet on the 'Society of Individuals.'" *Thesis Eleven* 94:29–48.

Brown, Patrick, and David Botstein. 1999. "Exploring the New World of the genome with DNA Microarrays". *Nature Genetic Supplement* 21 (January):33–37.

Chiang, Howard Hsueh-Hao. 2009. "The Laboratory Technology of Discrete Molecular Separation: The Historical Development of Gel Electrophoresis and the Material Epistemology of Biomolecular Science." *Journal of the History of Biology* 42:495–527.

Collins, Francis, Michael Morgan, and Aristides Patrinos. 2003. "The Human Genome Project: Lessons from Large-Scale Biology." *Science* 300 (11 April):286–290.

Consortium, The ENCODE Project. 2011. "A User's Guide to the Encyclopedia of DNA Elements (ENCODE)." *PLoS Biology* 9 (4):1–21.

Consortium, The ENCODE Project. 2012. "The Integrated Encyclopedia of DNA Elements in the Human Genome." *Nature* 489:57–74.

Consortium, International Human Genome Sequencing. 2001. "Initial Sequencing and Analysis of the Human Genome." *Nature* 409:860–921.

Consortium, International Human Genome Sequencing. 2004. "Finishing the Euchromatic Sequence of the Human Genome." *Nature* 431 (21):932–945.

Daston, Lorraine, and Peter Galison. 2007. *Objectivity.* Brooklyn: Zone Books.

Davis, Paula, James Kelley, Steven Caltrider, and Stephen Heinig. 2005. "ESTs Strumble at the Utility Threshold." *Nature Biotechnology* 23 (10):1227–1229.

Flax, Jane. 1993. *Disputed Subjects: Essays on Psychoanalysis, Politics, and Philosophy.* London: Routledge.

Funtowicz, Silvio, and Jerome Ravetz. 1992. "Three Types of Risk Assessment and the Emergence of Post Normal Science." In *Social Theories of Risk*, edited by Sheldon Krimsky and D. Golding. New York: Praeger.

Gall, Joseph. 1986. "Human Genome Sequencing." *Science* 233 (4771):1367–1368.

Gauchet, Marcel. 2000. "A New Age of Personality: An Essay on the Psychology of Our Times." *Thesis Eleven* 60:23–41.

Gauchet, Marcel. 2002. "Redefining the Unconscious." *Thesis Eleven* 71:4–23.

Gerstein, Mark, Can Bruce, Joel Rozowsky, Deyou Zheng, Jiang Du, Jan Korbel, Olof Emanuelsson, Zhengdong Zhang, Sherman Weissman, and Muichal Snyder. 2007. "What Is a Gene, Post-ENCODE? History and Updated Definition." *Genome Research* 17:669–689.

Green, Phil. 2002. "Whole-Genome Disassembly." *Proceedings of the National Academy of Sciences* 99 (7):4143–4144.

Green, Philip. 1997. "Against a Whole-Genome Shotgun." *Genome Research* 7:410–417.

Hacking, Ian. 2012. "Objectivity in Historical Perspective." *Metascience* 11:11–39.

Henikoff, Steven. 2007. "ENCODE and Our Very Busy Genome." *Nature Genetics* 39 (7):817–818.

Keller, Evelyn Fox. 2010. *The Mirage of a Space between Nature and Nurture.* Durham, NC: Duke University Press.

Kellis, Manolis, Barbara Wold, Michael Snyder, Bradley Bernstein, Anshul Kundaje, Georgi Marinov, Lucas Ward, Ewan Birney, Gregory Crawford, Job Dekker, Ian Dunham, Laura Elnistski, Eric Green, Roderic Guigo, Tim Hubbard, Jim Kent, Jason Lieb, Richard Myers, Michael Pazin, Bing Ren, John Stamatoyannopoulos, Zhiping Weing, Kevin White, and Ross Hardison. 2014. "Defining Functional DNA Elements in the Human Genome." *Proceedings of the National Academy of Sciences* 111 (17):6131–6138.

Kitcher, Philip. 1994. "Who's Afraid of the Human Genome Project?" *Proceedings of the Biennial Meeting of the Philosophy of Science Association* 2:313–321.

Lewin, Roger. 1986a. "Proposal to Sequence the Human Genome Stirs Debate." *Science* 232 (4758):1598–1600.

Lewin, Roger. 1986b. "Shifting Sentiments over Sequencing the Human Genome." *Science* 233 (4764):620–621.

Lewontin, Richard. 1992. "The Dream of the Human Genome." In *Biology as Ideology*, edited by Richard Lewontin. New York: Harper.

Maxam, Allan, and Walter Gilbert. 1977. "A New Method for Sequencing DNA." *Proceedings of National Academy of Science* 74 (2):560–564.

McElheny, Victor. 2010. *Drawing the Map of Life: Inside the Human Genome Project.* Philadelphia: A Merloyd Lawrence Books.

Moyn, Samuel. 2009. "The Assumption by Man of His Original Fracturing: Marcel Gauchet, Gladys Swain, and the History of the Self." *Modern Intellectual History* 6 (2):315–341.

Nowotny, Helga, Peter Scott, and Michael Gibbons. 2001. *Rethinking Science: Knowledge and the Public in an Age of Uncertainty.* Cambridge: Polity Press.

Olson, Maynard. 1993. "The Human Genome Project." *Proceedings of National Academy of Science of the United States of America* 90 (10):4338–4344.

Olson, Mynard, Leroy Hood, Charles Cantor, and David Botstein. 1989. "A Common Language for Physical Mapping of the Human Genome." *Science* 245 (4925):1434–1435.

Rose, Hilary, and Steven Rose. 2012. *Genes, Cells and Brains: The Promethean Promises of the New Biology.* London: Verso.

Rosenberg, Alex. 1994. "Sunversive Reflections on the Human Genome Project." *Proceedings of the Biennial Meeting of the Philosophy of Science Association* 2:329–335.

Rosenberg, Alex. 2006. *Darwinian Reductionism: How to Stop Worrying and Love Molecular Biology*. Chicago: University of Chicago Press.

Rowen, Lee, Gregory Mahiras, and Leroy Hood. 1997. "Sequencing the Human Genome." *Science* 278 (5338):605–607.

Sanger, Frederick. 1988. "Sequences, Sequences and Sequences." *Annual Review of Biochemistry* 57:1–28.

Shapin, Steven. 1992. "Discipline and Bounding: The History and Sociology of Science as Seen through the Externalism-Internalism Debate." *History of Science* 30:333–369.

Shreeve, James. 2004. *The Genome War: How Craig Venter Tried to Capture the Code of Life and Save the World*. New York: Ballantine Books.

Smith, David, and Robert Sinsheimer. 1986. "Human Genome Sequencing." *Science* 233 (4770):1246.

Söderqvist, Thomas. 1996. "Existential Projects and Existential Choice in Science: Science Biography as an Edifying Genre." In *Telling Lives in Science: Essays in Scientific Biography*, edited by Michael Shortland and Richard Yeo, 45–84. New York: Cambridge University Press.

Sulston, John, and Georgina Ferry. 2002. *The Common Thread: A Story of Science, Politics, Ethics, and the Human Genome*. Washington, DC: The Joseph Henry Press.

Swerdlow, Harold, and Raymond Gesteland. 1990. "Capillary Gel Electrophoresis for Rapid, High Resolution DNA Sequencing." *Nucleic Acids Research* 18 (6):1415–1419.

Tashima, Masaro, Bruno Calabretta, Giuseppe Torelli, Margaret Scofield, Abby Maizel, and Grady Saunders. 1981. "Presence of a Highly Repetitive and Widely Dispersed DNA Sequence in the Human Genome." *Proceedings of the National Academy of Sciences of the United Staes of America* 78 (3):1508–1512.

Tauber, Alfred, and Sahotra Sarkar. 1993. "The Ideology of the Human Genome Project." *Journal of the Royal Society of Medicine* 86:537–540.

Venter, J. Craig. 2007. *A Life Decoded: My Genome, My Life*. New York: Viking.

Venter, J. Craig, Mark Adams, Granger Sutton, Anthony Keravage, Hamilton Smith, and Michael Hunkapiller. 1998. "Shotgun Sequencing of the Human Genome." *Science* 280 (5369):1540–1542.

Venter, J. Craig, et al. 2001. "The Sequence of the Human Genome." *Science* 291 (5507):1304–1351.

Waterston, Robert, Eric Lander, and John Sulston. 2002. "On the Sequencing of the Human Genome." *Proceedings of National Academy of Science of the United States of America* 99 (6):3712–3716.

Waterston, Robert, Eric Lander, and John Sulston. 2003. "More on the Sequencing of the Human Genome." *Proceedings of the National Academy of Sciences* 100 (6):3022–3024.

Waterston, Robert, and John Sulston. 1998. "The Human Genome Project: Reaching the Finish Line." *Science* 282 (5386):53–54.

Weinstock, George. 2007. "ENCODE: More Genomic Empowerment." *Genome Research* 17:667–668.

Žižek, Slavoj. 2006. *The Parallax View*. Cambridge, MA: The MIT Press.

8

CONCLUSION

The history of science, including the history of genetic science and molecular biology, is commonly written as a chronological emergence of a series of ideas of individual scientists. For instance, the author of one of the popular science histories of molecular biology writes in his foreword to the first edition, "*The Eighth Day of Creation* is not a history of scientific ideas in abstract but of scientists in the process of discovery" (Judson 1979, loc 239). The book is a story of the discovery of DNA, RNA protein, but is based entirely on the conversations with more than 100 scientists probing into questions such as "who did what, and with what quirks of personality, and why they did it, and how they did it" (Judson 1979, loc 276). The history of science is often a series of discoveries made by "bright" and "extra-ordinary" people (Judson 1979, loc 327, 457). Even when the emphasis on brightness and extraordinariness, including the quirkiness of the personalities, might not be that evident in another form of history of molecular biology, the one that is organized as history of institutions – i.e., the history of synergy between intellectual capital and economic resources – as Lily Kay's wonderfully nuanced history of the new molecular biology in the post-war period does, it is inadvertently about the pioneering scientists and their practices in these institutions (Kay 1993, 2000). In a similar vein, while Keller declares that her primary focus in her excellent history of the gene concept over the twentieth century is to show the "ever-widening gap between our starting assumptions and the actual data", a great deal of her book chronologically discusses the unfolding of the gene concept as in the work of individual scientists (Keller 2000, 8). Whether the gene is discussed as a boundary object or "the concept in tension between its materiality and empirical instrumentality", the history of the gene is inevitably about a series of scientists and their ideas unfolding since the early twentieth century (Falk 2000, 318; Rheinberger 2000). These histories, however, remain mainly the history of abstract thought in which individual scientists often figure as carriers of ideas, whose presence plays a crucial but incidental role in the overall story. While these histories focus on an institutional, economic, and intellectual climate in which the science of the gene unfold, the knowledge claims remain disembodied and autonomous from the lives of individual scientists. Even the history of science that pays attention to scientists' subjectivity in the making of science, these are treated as mental states of collectives. The role of subjectivity of the individual scientist is often treated as mind reading, relegated to biographies and usually looked upon with suspicion as "individual reflexes" "that can wind up dissolving the subject-matter of history of science" into merely vision of a particular scientist (Shapin 1992, 354–355). Either way, the history of thought is resolutely divorced from the life experiences of scientists in the history of science.

In the recent texts in the history of science that look at the subjectivity-objectivity relationship, the scientist-subject is more than often theorized as neo-Kantian ideal – a unified and wilful, self-determined, self-regulated, active and autonomous, rational subject wilfully driven by social and scientific ethos – or a Foucauldian construct – reduced entirely to the effects of power. The psychoanalytical subject – the feeling, suffering, experiencing subject, the pathological subject split between the conscious and unconscious, the subject obstinately resistant to control by the conscious, the subject made of repressed drives that put serious obstacles in the path of the rational transformations – in short, the pathological subject of psychoanalysis is not part of this recent history of science. In a similar vein, Söderqvist discusses at length how the existential project of the individual scientist is not the subject of science biographies either and how treating the scientist-subject doing science as making existential choices of self-assertion and self-renewal can provide a different conception of science (Söderqvist 1996).

While this book is based on the challenge to the positivist and foundationalist concept, especially the neo-Kantian idea of subjectivity, it does not boast to provide an alternative normative or regulative theory of the scientist-subject. Such normative or regulative standard may turn into rigid structures while no such single regulative ideal can provide a prescription for human condition, suffering, or capacity (Flax 1993, 93). Subjectivity is conceived here as process expressing multiple and agentic qualities rather than as an effect of the predetermined normativity. The book thus derives from a variety of theoretical sources – philosophical, psychoanalytical, existential, and history of emotions and psychology of our times. It tells multiple stories in a variety of styles to appreciate the complexity of human and hence scientific subjectivity.

At the same time, the book is an attempt to reinterpret the history of reductionism in genetic science over the twentieth century as an affective/emotive history. Discussing select "contingent" moments in the history of genetic science that shaped the reductionist belief system, *I aim to posit in this book that scientists' particular ways of being and belonging pioneer the structures of rational and cognitive thought.* My claim is that intellectual paradigms are *affect worlds*, in other words, the conceptual theories are *isomorphic* with the world emotionally and existentially desired. The book aims to explain the power and dominance of reductionism in science, including the challenges to this dominance, by explaining how the method and philosophy of reductionism operates in the microcosm of the individual affective lives of scientists, and how in the creative struggles of scientists, reductionism is sustained, adopted, questioned, and challenged.

The main focus of Chapter 2 is to develop a detailed critique of the neo-Kantian idea of the scientist-subject popular among historians by closely engaging with the role of scientist-subjectivity in the making of objectivity in Lorraine Daston and Peter Galison's book *Objectivity*, and in Daston's later and earlier works "On Scientific Observation" and "The Moral Economy of Science" (Daston 1995, 2008; Daston and Galison 2007). I have posited four challenges to the neo-Kantian and Foucauldian constructions of the co-implication of psychology and epistemology presented in these texts. Firstly, following Jacques Lacan's work, I have argued that the subject of science constituted by the mode of modern science suffers from paranoia. It is not the fear of subjectivity interfering objectivity but the impossibility of knowing the truth of the *real* that causes paranoia. Here, I have argued that it is not ethos of objectivity that drives epistemology as Daston and Galison argue, but pathos of paranoia. The second challenge builds upon Kant's own denial that the perfect correspondence between the human will and the moral law is possible. Kant himself thought that an ethical human act is impossible without the component of "pathology". This questions Daston and Galison's argument that there is always ethical imperative at the core of epistemic virtue. The third challenge contests the way Daston and Galison take *appearance for being* in their application of the Foucauldian concept of *technologies of the self* in modelling the master scientist-self. The fourth challenge questions the notion of psychological and unconscious in the making of epistemology in Daston's later and earlier works. Against this background, I aim to make a claim that understanding and disclosing "entities" in the scientific domain presupposes an understanding of "being" in general. My aim in this chapter is to open up the discussion for an alternative conception of the scientist-subject and thereby an affective and existential formulation of science.

The remaining chapters discuss five pioneering moments in the history of reductionism in genetic science/molecular biology – the founding of

CONCLUSION

the idea of the particulate gene in the work of H. J. Muller, the gene as code-script as posited in the Nobel Prize–winning physicist Schrödinger's short book *What Is Life?*, the discovery of the double helix in the science of Rosalind Franklin, the challenge to the concept of the particulate gene in Barbara McClintock's theories on control and transposition, and the hyper-reductionism of Craig Venter in the sequencing and mapping of the human genome. Following is a summary of the co-articulation of the affective and the cognitive as discussed in each chapter.

Chapter 3 explains how in the first quarter of the twentieth century, the early attempts to understand the structure and replication of a gene in the reductionist terms were isomorphic with immortality ideologies. Hermann Muller was one of the first scientists who imagined and provided evidences to consider the gene as material reality – the gene as nothing but a physical and chemical unit that can be grinded in mortar and cooked in beaker. The reductionism during this period helped to make a subtle transition from "understanding" the impact of mutation on the structure of the gene to "controlling" human evolution through what Muller called positive eugenics. Muller's reductionism posited that knowing the realm of the microscopic world that the universe was given to us. This over-arching reductionist metaphysics was reflected both in his science and in his lifelong, unwavering belief in positive eugenics as a political tool to perfect human evolution. The reductionist theory of the particulate gene was absolutely prerequisite for the program of positive eugenics, and the obverse was true, too – the empirical results of his science, for example, the gene mutations induced by the application of X-ray, irrespective of their consequences, eventually strengthened his belief in the positive eugenics and thereby in the immortality of humanity. Muller's soul-belief in human immortality not only made his science always oriented to future possibilities, but it also provided a powerful searchlight for his research while he would have otherwise groped in the darkness of indeterminate empiricism.

Chapter 4 places the little book *What Is Life?* in the life of the Nobel Prize–winning physicist Erwin Schrödinger. The book *What Is Life?* made Schrödinger one of the founding fathers of molecular biology. It is often even projected as one of the most influential scientific books of the twentieth century. However, *What Is Life?* was not Schrödinger's only expression on the subject of hereditary material. All through his adult life he was fascinated by the questions of somatic and spiritual memory and its inheritance. This chapter discusses three other texts written before and after *What Is Life?* and argues that all through his adult life Schrödinger was concerned about the metaphysical search for unity and eternity of consciousness and permanence of memory, and *What Is Life?* was yet another and not the only expression

164

of this search. A great deal of historiography of *What Is Life?*, however, has reviewed and valued the book only for its contribution to reductionism in founding molecular biology and has ignored Schrödinger's philosophical and metaphysical thoughts so expressed in all four texts, including in the epilogue of *What Is Life?*. Not only that such history is presented in a progressivist fashion, but only those contributions count that add up to the "advancement" of science. Such history of science is not only committed to the concept of science as nothing but "radical empiricism", "epistemology without knowing subject" or "knowledge as disembodied ideas" (Amsterdamski 1975, 51; Söderqvist 1996), but it also portrays the scientist-self as a neo-Kantian ideal – he is foremost a rational individual making theory choices based purely on empirical evidences. Chapter 4 places *What Is Life?* in Schrödinger's life and argues that Schrödinger's emphatic reductionist declaration that the gene is nothing but physical and chemical entity in *What Is Life?* emerged from an entirely non-empirical and non-reductionist metaphysics. Considering Schrödinger's generative role, it could be argued that reductionism of structural molecular biology is rooted in the form of thought that the same science declares as irrational, discredited, and untenable.

Evelyn Fox Keller wrote the first biography of the Nobel Prize–winning geneticist Barbara McClintock in which Keller discussed how McClintock felt she was rejected by her peers in the 1950s because she questioned the dominant idea of the particulate gene and instead proposed that the genetic material jumped positions on the chromosome, which indicated that the gene did not control but was controlled by the cellular environment. Keller's story of McClintock's life is an account of a woman scientist's conception of science and how her unorthodox views isolated her from the mainstream science. Keller's biography was read by many in a way that made McClintock a feminist icon by showing how women scientists "see" scientific objects differently and how their science is holistic and hence radically different from the reductionism of male-dominated science. The second biographer, Nathaniel Comfort, calls this story a myth. In his detailed intellectual biography, Comfort embarks on an energetic journey to separate fact from fiction, to dismantle what he calls the McClintock myth. The difference between the two biographers is not entirely about evidences or about separating fact from fiction but about their adoption of two contrasting paradigms of a scientist's subjectivity: Keller foregrounds McClintock's affective self and Comfort her rational. In Chapter 5 I have closely and comparatively read both biographies to revisit Keller's "myth" and Comfort's "truth" and to provide yet another interpretation of McClintock's life and work from the perspective of object relations theories in psychoanalysis. Instead of figuring out the extent to which the myth bears truth, as Comfort does, I have asked questions: How and why this private myth was in the making throughout McClintock's life and work? How this private myth was related to the making of her science? By using a developmental psychoanalytical approach, I show that what Comfort calls McClintock's private myth was not something that was partly fictional and hence incorrect or wrong but it emerged from a deeply and compellingly affective place in McClintock's life. This so-called myth was integral to and fundamentally formative of who she was, a woman and a scientist, and this myth formatively shaped McClintock's relationship with science's objects and science's subjects. In fact, I show how in terms of Ian Hacking referring to Kant, objectivity in McClintock's science was "super-duper intersubjectivity".

Rosalind Franklin is immortalized as an unsung hero or a victim of discriminatory practices in the debates on the gender and accreditation of science. In Chapter 6 I contest that such representation of the epistemic agent not only refers to science as a product but implied in such images is a caricature of the scientist as an autonomous, self-determining, and self-regulated individual whose science is nothing but a result of rational cognitive choices. In revising the notion of the scientist-self as fundamentally feeling, suffering, and experiencing – the decentred-subject – and in engaging with the science of the double helix as a process, Chapter 6 traces the agency of Rosalind Franklin unfolding on the interface of her situatedness, her sexuality, and her existential self-assertion in intra-action with not only other epistemic agents but with the materiality of epistemic apparatuses and concepts. In doing so, I challenge the simplistic images of Franklin as either a victim or a hero and show that Franklin's agency in the making-of-the-science of the double helix was fraught with contradictions in such a way that she occupied positions of domination *and* marginality at the same time during different temporal and epistemic locations in the process. Franklin's science-in-the-making also shows that the methodological choices she made were hardly the conscious, deliberate choices usually ascribed to a rational and calculating scientist-agent in pursuit of "truth". In this chapter I show how the existential desire of self-assertion was the motivational force behind Franklin's science of the macromolecule.

Craig Venter, widely considered a brilliant but maverick scientist, working with a private company in 1998, declared that he intended to complete sequencing the human genome four years in advance of the finish line set by the state-funded public consortium. This started a "race", also often described as "war", on the human genome project between the public and private scientists. On his own frequent admission, the human genome project for Venter was first and foremost about "winning" and "speed" and only secondarily about science and making money. Chapter 7 looks at the way in which Venter's affective self, which following the French philosopher Marcel Gauchet's work I call hyper-contemporary personality, was consistent or isomorphic with the cognitive choices he made and how these choices, on the one hand, forced the direction and logic of the public consortium, and on the other, resulted into hyper-reductionism of human genome science.

Based on this detailed discussion on the life and work of each scientist, I wish to highlight two conclusions. Firstly, the book shows how the choice of the method of reductionism does not mean that the scientist-subject is selecting from an infinite list of possibilities (as if choosing a product in the supermarket) but chooses the "necessities that determine itself". For Žižek, this is the Lacanian paradox of the subject, a parallax moment, the choice that is no choice, or the choice that is nothing but choosing the necessity of one's own existence and hence no choice (Žižek 2006). The chapters also show how in the lives of individual scientists the making of objective science was not only an interplay between the rational and irrational but also the intertwining of the intrapsychic and intersubjective worlds. Secondly, the book makes a case for an alternative history of reductionism in genetic science by showing the way in which the method and philosophy of reductionism was challenged, questioned, or adopted in the microcosm of the individual lives of scientists. In each specific case, it shows not only how the *existence precedes essence* meaning the deeper affective drives pioneer ethos in science, but the chapters also show that not only the connection between the existence and essence is resiliently robust but this connection between the affective and the cognitive is highly specific and compelling in shaping the life and work of each one of them. Not only that the affective provided powerful motivational force for the choice of empirical problem, but that it has been profoundly constitutive of the making of the science itself. The affective history of the gene has further consequences for the history of method in science and the history of epistemology, however, these remain to be explored.

References

Amsterdamski, Stefan. 1975. *Between Experience and Metaphysics: Philosophical Problem of the Evolution of Science.* Dordrecht-Holland: D. Reidel Publishing Company.
Daston, Lorraine. 1995. "The Moral Economy of Science." *Osiris* 10 (2nd Series):2–24.
Daston, Lorraine. 2008. "On Scientific Observation." *Isis* 99 (1):97–110.

Daston, Lorraine, and Peter Galison. 2007. *Objectivity*. Brooklyn: Zone Books.

Falk, Raphael. 2000. "The Gene: A Concept in Tension." In *The Concept of the Gene in Development and Evolution: Historical and Epistemological Perspectives*, edited by Peter Beutron, Raphael Falk and Hans-Jorg Rheinberger, 317–348. Cambridge: Cambridge University Press.

Flax, Jane. 1993. *Disputed Subjects: Essays on Psychoanalysis, Politics, and Philosophy*. London: Routledge.

Judson, Horace. 1979. *The Eighth Day of Creation: Makers of the Revolution in Biology*. New York: Simon & Schuster.

Kay, Lily. 1993. *The Molecular Vision of Life: Caltech, The Rockefeller Foundation, and the Rise of the New Biology*. Oxford: Oxford University Press.

Kay, Lily. 2000. *Who Wrote the Book of Life? A History of the Genetic Code*. Stanford, CA: Stanford University Press.

Keller, Evelyn Fox. 2000. *The Century of the Gene*. Cambridge, MA: Harvard University Press.

Rheinberger, Hans-Jorg. 2000. "Gene Concepts: Fragments from the Perspectives of Molecular Biology." In *The Concept of the Gene in Development and Evolution: Historical and Epistemological Perspectives*, edited by Peter Beutron, Raphael Falk and Hans-Jorg Rheinberger, 219–239. Cambridge: Cambridge University Press.

Shapin, Steven. 1992. "Discipline and Bounding: The History and Sociology of Science as Seen through the Externalism-Internalism Debate." *History of Science* 30:333–369.

Söderqvist, Thomas. 1996. "Existential Projects and Existential Choice in Science: Science Biography as an Edifying Genre." In *Telling Lives in Science: Essays in Scientific Biography*, edited by Michael Shortland and Richard Yeo, 45–84. New York: Cambridge University Press.

Žižek, Slavoj. 2006. *The Parallax View*. Cambridge, MA: The MIT Press.

INDEX